# METHODS AND EXPERIENCES IN IMPACT ASSESSMENT

# METHODS AND EXPERIENCES IN IMPACT ASSESSMENT

Edited by

**Henk A. Becker**

Research Unit Planning and Policy Making,
Department of Sociology,
Rijksuniversiteit Utrecht, The Netherlands

and

**Alan L. Porter**

Technology Policy and Assessment Center,
Georgia Institute of Technology,
Atlanta, U.S.A.

Reprinted from
Impact Assessment Bulletin, Vol. 4 No. 3/4

D. Reidel Publishing Company

A MEMBER OF THE KLUWER ACADEMIC PUBLISHERS GROUP

Dordrecht / Boston / Lancaster / Tokyo

**Library of Congress Cataloging in Publication Data**

CIP

Methods and experiences in impact assessment.

"Offspring of the fourth conference of the International Association for Impact Assessment, held in Utrecht, the Netherlands, from 27 to 28 June 1985" – Editorial note.

Includes index.

1. Technology assessment–Congresses. I. Becker, H. A. II. Porter, Alan L. III. International Association for Impact Assessment.

T174.5M48 1986 363.1 86-10043

ISBN 90-277-2264-1

---

Published by D. Reidel Publishing Company
P.O. Box 17, 3300 AA Dordrecht, Holland

Sold and distributed in the U.S.A. and Canada
by Kluwer Academic Publishers,
190 Old Derby Street, Hingham, MA 02043, U.S.A.

In all other countries, sold and distributed
by Kluwer Academic Publishers Group,
P.O. Box 322, 3300 AH Dordrecht, Holland



Printed in The Netherlands

TABLE OF CONTENTS

EDITORIAL NOTE

This volume is an offspring of the fourth conference of the International Association for Impact Assessment. IAIA held this meeeting in Utrecht, the Netherlands, from 27 to 28 June 1985. The editorial introduction provides the reader with an eye witness account of the conference. The introductory chapter also specifies where a second selection of papers will be published.

The main theme of the conference was "methods and experiences in impact assessment". Bearing this theme in mind the editors brought together in this volume papers that deal with the "state of the art" in the field of methods and that provide strategic examples of the work that is being done in impact assessment in a variety of countries.

The editors want to express the gratitude of IAIA for the financial support the association has received for the publication of these proceedings. The Division for Science Policy of the Department of Education and Science in The Netherlands has been most generous in providing all the support that was needed.

The editors also want to thank Marcel Reijnen for his competent and dedicated assistance and the secretarial staff of the Research Unit for Planning and Policymaking, State University at Utrecht, for their patient and efficient activities.

The editors:

Henk A. Becker
Alan L. Porter

# IMPACT ASSESSMENT, FROM UTRECHT

Alan L. Porter*

I wish to highlight the Conference as context from which this book emerges. This Fourth Annual Conference of the International Association for Impact Assessment is the first held in Europe. For two days (June 27-28, 1985) some 170 participants gathered to consider environmental, risk, social, and technology assessment issues at Utrecht in The Netherlands.

The conference was hosted by Utrecht University, founded in 1636 a few months before Harvard, yet only two years into the experience of a dramatic reorganization to direct multiple disciplines to work together on specific problems. At last count, 69 well-written papers were in hand to counter-balance against a strict page limit for this volume. Applying multiple criteria of freshness of ideas, quality of research, and topical distribution led to the book before you. A second volume, entitled Impact Assessment Today, is in preparation by the same editors, to be published by van Arkel (The Netherlands).

Four special speakers oriented the conference participants toward the big issues before us. I would first like to share some of their thoughts for they bear on what impact assessment is, and can become.

Our special guest speaker, Dr. Piet Thoenes, addressed the state of the Western European welfare state, and planning and citizen participation therein.

---

* Georgia Institute of Technology, Atlanta, Georgia

Recent times have found growing dissatisfaction with the Welfare State, both by those who give and those who take. "... the Welfare State bureaucracy was ideologically no longer trustworthy, that as an instrument it was too clumsy and too slow to give due honours to the newly developed sense of justice and fairness. Within the society arose new divisions: workers and non-workers; men and women; young and old; born here, born elsewhere. New divisions breaking up the Welfare State solidarity".

With this splintering of Welfare State unity in countries like Denmark and The Netherlands, came increasing demands for a say in what government did. "Fundamental democratisation is one of the essential characteristics of western societies." After 1960, this led to a wave of "inspraak en medezeggenschap" (public participation) in the family, in the factory, in the school, in church and state. "After 1970 however things went really wrong. The economy deteriorated and the government as a large scale entrepreneur, rescuing weak firms and stimulating employment, did not show up as an able long term planner"..... "Anyway in recent years there are examples galore of planning which failed because of an overdose of participation and of participation which failed because of planning with an overdose of ununderstandable technicalities."

Given difficulties with technocratic planning and with fully democratic participation, what shall we do? Dr. Thoenes takes acid rain as an example. "Acid rain and many other of our ecological miseries do not stop at political boundaries. The whole affair can be started and pushed by idealistic individuals and protest movements. However, if we want to solve our problem we should find the courage to direct matters to our governments and to let them make a long term plan for action. More so we should let the EEG make a plan for the whole of Western Europe".

Dr. Thoenes closes with two French quotes:

"Pour les grands maux, les grandes rémèdes. (For big

problems, big solutions).

and from Danton, what he said in a period when France was making world history:

"De l'audace, de l'audace et encore de l'audace". (Courage, courage, and evermore courage). That is what we should keep in mind today in a period where it is necessary to plan in uncertainty."

Still within the context of planning, but moving to impact assessment concerns per se, Dr. E. van Spiegel presented the perspective of Mr W.J. Deetman, Minister of Education and Science for The Netherlands. Of particular interest are descriptions of Dutch programs for technology assessment and for research in technology assessment.

Noting everyone's emphasis on technological advances, Mr. Deetman asserts that new technology "may also have unexpected and unwelcome side-effects which may threaten our quality of life. In recent years attention has been increasingly devoted to these unexpected and unintentional results of technological progress."

Technology assessment has a role to play in policy making. "I would argue that policy-makers are increasingly coming to realize that an understanding of all significant effects of a policy is a precondition of good management, optimal deployment of staff and resources and proper accountability, and that this knowledge should ideally be present at an early stage of the proceedings. Even where Parliament does not make demands in this regard, information is being collected on the impact of policies; thus far, however, this has not been done in a sufficiently systematic way ... I am convinced that research to determine the impact of policies, whether or not it is initiated by the government, can help to improve the relationship between the authorities and the public and to legitimate policy. It is important, however, that research of this kind should not be seen as an instrument of central bureaucratic con-

trol.
(He continues) ... I (Mr. Deetman) presented a policy document to Parliament last year on integrating science and technology into society. The document was based on discussions in Parliament over many years on a number of subjects, the most important of which were recombinant DNA research, micro-electronics, innovation, energy and the environment. When the present Government was formed, I was instructed, in my capacity as coordinating Minister for Science Policy, to take steps to increase public understanding of the social and ethical consequences of scientific and technological developments. I decided to approach this issue by improving the decision making process in the field of science and technology. Efforts would therefore be made to anticipate possible effects of new advances more promptly, and more carefully and to ensure closer involvement by all those concerned. Underlying these measures was the belief that the quality and the effectiveness of new technology are not determined exclusively by the nature of the innovations themselves. The way in which account is taken of ethical and social consequences when such innovations are introduced is of equal importance..."

It was explicitly decided to make public information a major element in the program designed to integrate science and technology into society. An independent organization will be set up for this purpose, in conjunction with the Ministry of Economic Affairs, which is responsible for coordinating technology policy. Among the new organization's most important tasks will be to support information services on science and technology as a whole, to carry out joint projects with other organizations and to devise additional or innovative information activities of its own....

(A technology assessment studies research program) would involve putting plans for a research program before Parliament each year, based on a systematic analysis of suggestions made by Parliament, government departments, research institutes and interest groups. As regards the last of these, the Government intends to experiment with what we provisionally call the "public post box" which would also operate on a decentralized basis with the help of existing organizations, to deal with reactions from the public at large.

"...the program should act as an impulse towards innovation by, for instance, concentrating on technology assessment studies which examine hitherto neglected aspects of the question or the impact of technology on groups ignored thus far, and research focusing on new developments such as information technology or biotechnology or on developing and testing a methodology of technology assessment."

In addition to Mr. Deetman's innovative spirit in technology assessment, I was impressed by a sensitivity to building social science data infrastructure. "In the interests of improved cooperation, a national network is being set up for data and field work facilities and an advisory committee is currently engaged in drawing up plans." Perhaps other nations, including the United States, could profitably follow these Dutch initiatives in technology assessment.

Having considered global and national policy contexts for impact assessment, we follow the Dean of the Faculty of Social Sciences at Utrecht, Dr. J.H. Dijkhuis, into the issue of who should be involved.

"Under the aegis of the social disciplines, in particular sociology, impact assessment has addressed itself first and foremost to the social consequences of interventions. As a result of this orientation, it is noticeable that more and more attention is being paid to the experience of change, and the stresses it gives rise to. Looking across to the behavioural sciences, in particular my own discipline of psychology, one cannot help noticing the increasing amount of attention being paid to the effect on behaviour of (changing) social conditions. Thus, in psychological research on stress, what has come to occupy the centre of the stage are the external causes of stress, and the social support systems which can alleviate it. The study of what happens when changes are introduced in human situations must be pre-eminently a multidisciplinary task, because the different levels (meso, macro, micro) form, as it were, inseparable links in a chain."

Dean Dijkhuis goes on to note impact assessment work at Utrecht. "The first publication appeared in 1969. It dealt with the impact of an increase in the number of academics in the managerial labour market in The Netherlands. Subsequently, projects were initiated on regional development and on the future of our country as a whole. These studies utilized scenario techniques. More recently, projects have been developed in the area of health care systems, with particular reference to the problem of aging. The impact of innovation in educational systems has also been investigated. Lastly, I would like to mention the attention which is devoted in our faculty to the study of research methods, in particular scenario techniques. The Board of the Faculty intends to do everything in its power to stimulate further research on the field of impact assessment."

A fitting close to this preamble was offered by the Rector Magnificus of the University, Dr. Otto J. de Jong, quoting the New Testament. "Would any of you think of building a tower without first sitting down and calculating the cost, to see whether he could afford to finish it? Otherwise, if he has laid its foundation and then is not able to complete it, all the onlookers will laugh at him: ´There is the man´, they will say, who started to build and could not finish´."

My second purpose in this introduction is to outline what is to come. Table 1 offers my "instant index" - my perception of which papers address which topics. Not surprisingly, many papers touch on a range of topics. That makes it difficult to order the papers in a definitive fashion. However, the columns show the papers in the order they appear and there are clusterings. The first three papers explicitly address the future; they do forecasting. The next four papers (Murdock et al. through Muth and Lee) emphasize sociale impacts. Barendregt and Walker deal with water, with Walker, Nijkamp, and Papp each covering some aspects of impact evaluation and policy analysis. Canter and Fairchild, Carly, and Storey go beyond impact assessment to develop post-EIA impact management. Leyten compares how five Western European countries are organizing technology assessment efforts. And, the final five papers (Evers through Waller) direct attention to impact assessment in developing countries.

Table 1 also provides a quick scan of current issues in impact assessment (IA). In find the focus on building theoretical foundations for social impact assessment (SIA) notable. Likewise, note the attention directed beyond IA per se to monitoring of actual impacts sustained from developments, to evaluate how well IAs have anticipated these, and to build impact management programs. As the tally of case settings shows, the conference presented a good blend of European and North American experiences. It also points to the "growth" area of IA in developing countries.

Following are brief synopses of the papers to aid you in identifying those of keen interest.

Becker describes the construction of health care scenarios incorporating major health developments influencing The Netherlands to 2000, and identifying important alternative health service patterns to be considered.

Keilman concentrates on how to quantify uncertainty in population forecasting, considering variants due to the effect of qualitative social hypotheses affecting quantitative trends. Forecasting practice and errors in The Netherlands serve as illustration.

Rossini forecasts office automation technology developments through 2000, an arena in which changes in information technologies will drastically affect work life.

Murdock et al. assess the present status of socio-economic impact assessment, examining the state of knowledge in predicting economic, demographic, public service, fiscal and social impacts. They spell out needs for conceptual and methodological improvement and offer three scenarios for the future of socio-economic IA.

Boskma presents evidence that social simulation can allow realistic social interaction under simulator control. This can contribute to SIA, with certain validity limitations discussed.

Peters describes the use of planning cells, with citizens paid to participate, to evaluate four alternative energy future scenarios for West Germany. The use of value tree analysis provides a vehicle to arrange expressed values in a hierarchical structure.

Muth and Lee present a compelling argument for considering social structures as explanatory, rather than simply descriptive, factors in SIA. They exemplify this in an empirical study of structural differentiation in 52 rural Alaskan communities as a predictor of social pathology.

De Boer develops a model of parallel social processes at work behind community response to soil pollution, backed by extensive interview data from The Netherlands.

Walker presents a systematic method for screening candidate policy alternatives to determine the most suitable set on which to conduct IA. Screening is illustrated in a study to develop water management policy for The Netherlands.

Nijkamp reviews multiple criteria analysis methods for applicability to impact evaluation.

Canter and Fairchild lay out a systematic approach for post-EIS (environmental impact statement) environmental monitoring.

Carley conceptualizes post-IA socio-economic and environmental monitoring in eight categories pertinent to policy processes.

The case of an offshore oil and gas development in Canada. Social impact monitoring is presented as an important tool in the overall policy process.

Evers discusses how the Organisation for Economic Co-operation and Development (OECD) has attended to EIA for aid projects to developing countries. Ways to prioritize those projects most in need of assessment and identification of constraints in carrying out EIA warrant attention.

Porter et al. construct a technology delivery system model to assess the potential uses and implications of microcomputers in developing countries.

TABLE "INSTANT INDEX"

| ARTICLE FEATURES | Becker | Keilman | Rossini | Murdock et al | Boskma | Peters | Muth & Lee | De Boer | Walker | Nijkamp | Canter & Fairchild | Carley | Evers | Porter et al |
|---|---|---|---|---|---|---|---|---|---|---|---|---|---|---|
| TOPICAL FOCUS | | | | | | | | | | | | | | |
| ENERGY | | | | v | | v | | | | | | | | |
| BIO/HEALTH/TOXIC WASTE | v | | | | | | | v | | | | | | |
| WATER | | | | | | | | | v | | | | | |
| INFORMATION TECHNOLOGY | | | v | | | | | | | | | | | v |
| FORECASTING | | | | | | | | | | | | | | |
| SOCIAL/DEMOGRAPHIC/SCENARIOS | v | v | | | | v | | | | | | | | |
| TECHNOLOGICAL | | | v | | | | | | | | | | | v |
| IMPACT ASSESSMENT | | | | | | | | | | | | | | |
| ENVIRONMENTAL-- (METHODS) | | | | | | | | | | | | | v | |
| SOCIAL -- (METHODS) | | | | v | v | v | | | | | | | | |
| SOCIAL PROCESSES/SIA THEORY | | | | v | | | v | v | | | | | | |
| SOCIO ECONOMIC | | | | | | | | | | | | | | v |
| IMPACT EVALUATION/POLICY ANALYSIS | | | | | | | | | | | | | | |
| SELECTION OF ALTERNATIVES FOR IA | | | | | | | | | v | v | | | v | |
| PRIORITIZING ALTERNATIVES-- METHODS, PARTICIPATION | | | | | | v | | | | v | | | | |
| POLICY PROCESSES | | | | | | | | | | | | v | | v |
| MONITORING/EVALUATION OF IA! | | | | v | | | | | | | v | v | v | |
| ORGANIZING FOR IA | | | | | | | | | | | | | v | |
| CASE SETTINGS | | | | | | | | | | | | | | |
| WESTERN EUROPE | v | v | | | | v | | v | v | | | | v | |
| NORTH AMERICA | | | v | v | | | v | | | | v | v | | |
| DEVELOPING COUNTRIES | | | | | | | | | | | | | v | v |

## PART A : FUTURES RESEARCH

A1 : Becker, H.A., Klaassen-van den Berg Jeths, Kraan-Jetten, A., Rijsselt, R. van, Contextual Scenarios on the Elderly and their Health in the Netherlands

A2 : Keilman, N., The Unpredictability of Population Trends

A3 : Rossini, F.A. and Porter, A.L., A Forecast of Office Automation Technologies through 2000

# CONTEXTUAL SCENARIOS ON THE ELDERLY AND THEIR HEALTH IN THE NETHERLANDS

Dr. Henk A. Becker *
Anneke Klaassen - van den Berg Jeths, M.A.
Andrea Kraan - Jetten, M.A.
René van Rijsselt, M.A.

ABSTRACT

This paper focuses on autonomous developments confronting policymakers responsible for health care for the elderly in The Netherlands. The autonomous developments are grouped into three contextual scenarios. Two "critical incidents" are assessed as to their potential impact on the health and social situation of the elderly. The research project described in this paper aimed at creating a "learning environment" for pre-testing the consequences of health policy measures for the elderly in The Netherlands.

KEY WORDS

health care for the elderly, contextual scenarios, critical incidents, learning environment, pre-testing the consequences of health policy measures

---

* Research Unit Planning and Policymaking, State University of Utrecht

## 1. THE STRUCTURE OF THIS PAPER

This paper wants:

- to tell something about the experiences in our scenario project;
- to give some results of the project;
- to deal with the prospects for using the results

In our project we assessed the impact of a number of "autonomous" developments and a number of "critical incidents" on patterns of health care in the years to come. Our project also provides a framework for assessing the impact of policy activities proposed in the Government Memorandum on "Health Care 2000", that will be issued in the near future.

Our project was financed by a Royal Commission on Futures Scenarios in Health care, installed by the Minister of Welfare, Health and Culture in The Netherlands. This Commission (STG) has also financed scenario projects on cancers, cardiovascular diseases and lifestyles. In 1985 the STG has launched a number of new projects.

The question that had to be answered in our project can be summarized as follows:

- what are the major (future) developments that will influence the health status of the elderly in The Netherlands between 1984 and 2000 ? and

- what are the most important alternative patterns of health services between 1984 and 2000, taking into consideration the future health status of the elderly and their relative number in de population in The Netherlands ?

The project for scenarios on aging and the elderly was carried out by the Scenario Commission on Aging, a working group of experts in the field of medical and related research, health services and health policies. The Scenario Commission was chaired by Professor Hollander. A research team provided methodological and empirical assistance. The research team provided methodological and empirical assistance. The research team consisted of Professor dr. H.A. Becker, project coordinator, mrs A. Klaassen, M.A., of the Research Unit in Planning and Policy Making, State University of Utrecht, The Netherlands.

The final report, called "Ouder worden in de toekomst" will be published in September 1985. The translation in English is in preparation. The date of publishing is probably at the end of 1985.

## 2. THE APPROACH CHOSEN IN THE PROJECT

### 2.1. problem analysis

To what extent will the elderly constitute a problem in society between 1984 and 2000 ? At the outset we have to keep in mind that, just as in the last decades, the majority of the elderly will enjoy an active life, will be relatively healthy and will live in their own homes. Without doubt the problems related to the elderly will be severe between 1984 and 2000, but nothing is to be gained by exaggerating them.

In our problem analysis we decided to define "the elderly" as all citizens of 55 or older. This decision is based on the assumption that at about the age of 55 the first signs of social disengagement related to getting older become apparent. The decision is also based on the assumption that illnesses and disabilities that affect the health status of people in old age will develop or become aggravated in the mid fifties. Of course, this decision with regard to the lower limit of the category of "the elderly" is an arbitrary one. Within the category of the elderly we distinguish between the "young elderly (55-64), the "old" (65-79) and the "very old" (80 +).

Taking 55 as the lower limit of the category implies a lot of the elderly involved will still have jobs. In our analysis, employment and other kinds of productive work are regarded by many of the elderly as an essential part of their existence. We assume that this will also be the case in the next decades. We even take into condideration the possibility that in the coming decades, more of the elderly than was the case between 1970 and 1984 will be engaged in jobs and in productive activities of a similar nature. in fields where expertise is scarce, the still capable elderly will be in strong demand.

During our problem analysis we were also confronted with the necessity to take interlinkages between medical, economic and social processes into consideration. Improving the quality of care for instance requires a joint effort.

## 2.2. baseline analysis 1970-1984

Which part of the past has to be taken into consideration if we want to look at developments up to the year 2000, and a little bit beyond ? We decided to take 1970 as a starting point. Between 1968 and 1973 a "cultural shock" was felt in The Netherlands: the welfare state was challenged and the first signs of the economic recession were appearing. To those who are 55 now, or will become 55 between now and 2000, the beginning of the seventies was a landmark in a social and an economic sense. But of course the changes are only relative.

The baseline analysis in this project was carried out by studying documents (research reports, governmental documents, theoretical publications). It also involved a number of quantitative analyses on statistical data.

The scheme of variables handled in the baseline analysis will be dealt with later on in this paper. The clusters of variables involved are indicators of developments that have taken place, and that might continue in the coming years.

The baseline analysis carried out in this project is closely related to the questions that had to be answered in this project. We had to focus and select quite rigorously. We present the results in such a way, however, that readers interested in a baseline analysis of a somewhat different kind can develop their own baseline analysis, taking our analysis as a starting point. Adjustment of the baseline analysis may be required for instance, if readers want to design scenarios that differ from our scenarios. The baseline analysis is "user friendly" enough to permit adjustments.

## 2.3. scenarios on autonomous developments 1980 - 2000

As a time horizon for our scenario project, we took the year 2000. Some developments, for instance demographic trends, are looked at up to the year 2030. Decisions like these cannot be but arbitrary.

Our scenarios are approximations of developments that are to a high degree "autonomous" if looked at from the position of individuals and organisations that are responsible for the policies concerning the health of the elderly in The Netherlands. These "focal actors" in our analysis have little or no power to change these developments. The autonomous developments involved are as a rule international changes affecting all countries, including small ones like The Netherlands.

The three contextual scenarios developed in our project are based upon forecasts, explorations and speculations we found in the literature. They are also based on the outcomes of discussions we had with groups of experts in the field of medical, biological and technological research and practice.

The selection of these three scenarios has to be explained, of course. We will give arguments in our final report, but ultimately the choice is to a certain extent an arbitrary one. This is no serious handicap, however. because the scenarios are presented in such a way that the reader can construct his own additional scenarios with not too much effort.

2.4. contextual scenarios as a "learning environment" for developing strategies

By taking (more or less) autonomous developments in the medical, economic, social and other fields as the "environment" for the policy making by the "focal actors", the scenario team follows a tradition that is quite strong in futures research nowadays. Reliable and precise forecasts of medical, economic and social developments are feasible on a relatively short term only. One to three years is often a maximum time span for reliable and precise forecasts. Demographic trends are relatively predictable: for instance the total number of elderly persons in The Netherlands can be predicted for a period of about 50 years. When forecasts are impossible, scenarios can help us in exploring the future. They sketch potential "contexts" for decision making with regard to the future. These "contexts" provide policy makers with a "learning environment". They can "test" the strategies they are considering in order to answer the question as to the circumstances in which these strategies would be feasible.

## 3. PITFALLS ENCOUNTERED

During our project we encountered a lot of pitfalls. We describe the most important of them here. This description may be useful for those who want to apply the results of our scenario project, as they illustrate the limits to the validity of our outcomes.

3.1. decisions with regard to the scope of the project

After long deliberation we decided to look not only at the health status of the elderly but also at health services, developments in medical, biological and technological fields and societal developments (economic, social) as well. We have already pointed out that choosing 55 years as the lower age limit for the analysis was an arbitrary decision.

### 3.2. coping with risk and uncertainty

To what extent is it possible to come forward with valid and precise information with regard to the future of aging and the elderly ? Looking at autonomous developments we distinguish between processes that show

a) high probability, high impact;
b) high probability, low impact;
c) low probability, high impact;
d) low probability, low impact.

Most processes registered in the contextual scenarios are either "high probability-high impact" or "high probability-low impact". To provide a learning environment for designing strategies which would be as realistic as possible, we added some processes that are characterized as "low probability-high impact". These are called "critical incidents". (Sections 7 and 8). Processes showing "low probability-low impact" are disregarded in our analysis.

### 3.3. selecting time horizons

Most developments involved can in the long run only be dealt with in an exploratory or even speculative way. Demografic processes as an exception allow for "forecasts" even up to 2030. The Netherlands Central Bureau of Statistics issued forecasts of demographic developments from 1980 to 2030 [1]. This bureau presented a low, a medium, and a high forecast. Variations are to be expected due to fluctuations in fertility, international migration and the number of marriages. In our analysis we took the middle alternative as a starting point. We are aware of the fact that the lower alternative is nowadays considered by many experts as the most valid forecast. However, the category of the elderly (55+) is hardly affected by these fluctuations, if we look at absolute numbers. If we look at the size of the category of the elderly in relation to the total Dutch population, there is some difference between the three alternatives.

### 3.4. "autonomous" developments and their relative rigidity

An important guess we had to make regarding the future involves the development of medical knowledge and know how. How likely are breakthroughs in the medical sciences in the next fifteen to twenty years? Would they substantially affect the health status of the elderly? According to one opinion, the medical sciences have harvested the benefits from innovations with regard to infectious diseases. At the moment there is not enough evidence to predict that major breakthroughs may be expeçted within the next twenty years. Innovations on a large scale require many years of testing on a small scale before they can be used on a large scale. According to other opinions medical breakthroughs are less unlikely.

In our project we treated the prospect of major medical innovations in the next fifteen to twenty years as "low probability-high impact". In some specific fields the prospects are less unfavourable. We explore the character and scope of one potential breakthrough: the postponement of the chronic brain syndrome by five years (Section 7).

Another pitfall is the approach to the course of future social developments. We explore the character and the impact of a decline in human solidarity (Section 8). A development like this would affect the health status and health care of the elderly in an important way. Changes in solidarity patterns in modern society are primarily regarded as autonomous processes.

### 3.5. the potential impact of alternative strategies

Can human solidarity be increased by planned change ? Can lifestyles be changed in a systematic way? A number of problems related to the future of the elderly demand social change. What is " the state of the art" in planned social change?

In the sixties in the United States and in Western Europe, planned social change was believed to be possible on a large scale. Experiences since then with planned changes have often been quite negative. Some social scientists have become pessimistic. Others have analyzed positive examples of planned change.The positive examples have a number of characteristics in common. As a rule they are related to changes that are already taking place. They speed up or channel these changes. Another characteristic is that they require more time and money than is usually allotted to projects of planned change.

## 4. VARIABLES AND OUTCOMES OF THE BASELINE ANALYSIS

In figure 1 the clusters of variables that are dealt with in the baseline analysis are described. This baseline analysis is not an integrated empirical model. Its primary objective is to focus the analysis and to avoid omissions and contradictons in the description. A schema of variables like this has been called a "consistency machine". This term indicates precisely where the principle task of a "chart of arrows" in our analysis lies (see Figure 1).

We shall not attempt to summarize the results of the baseline analysis in this paper. In the following sections we shall only give a short overview of the three most important variables of this analysis.

### 4.1. demographic developments

In 1983 The Netherlands had about 14.5 million

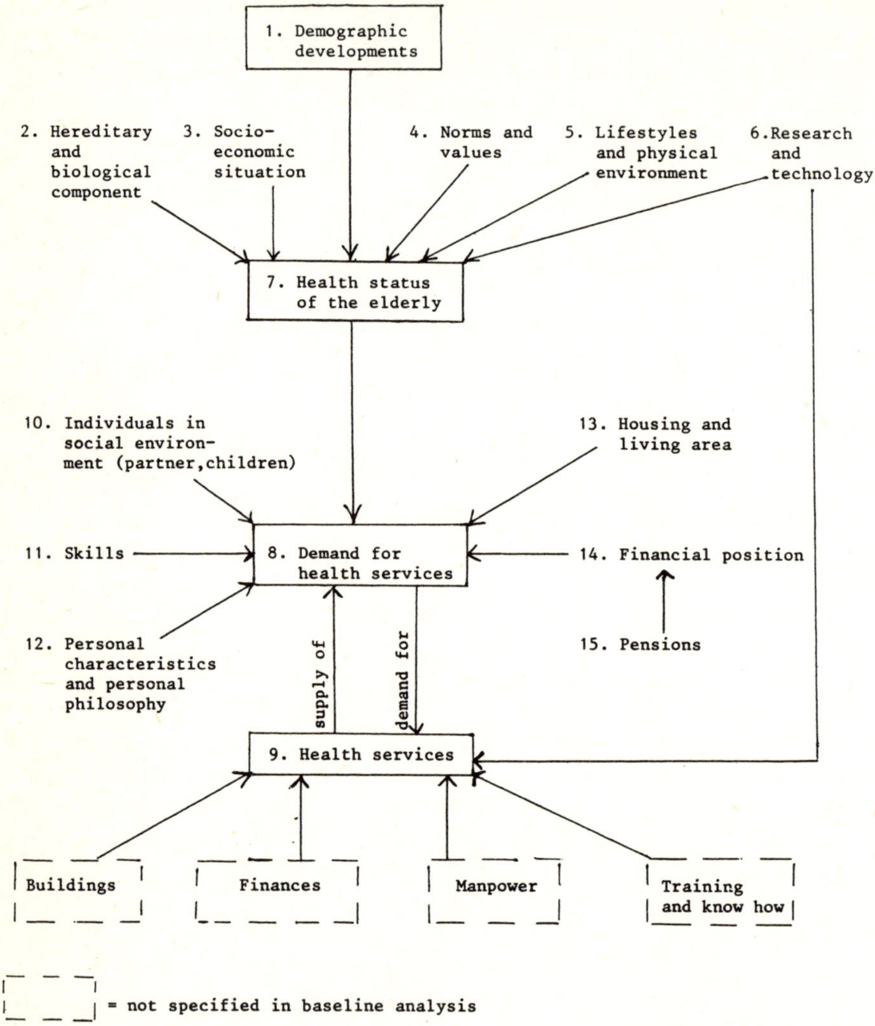


Figure 1: Clusters of variables in the baseline analysis

inhabitants. 3 million were older than 55 (21%) and

1.7 million were older than 65 (11.8%). According to the middle variant[2] there will be in the year 2000 about 3.7 million inhabitants of 55 years and older (23.4%) and 2.1 million inhabitants of 65 years and older (13.4%). These numbers will increase: in 2030 about 5.4 million inhabitants will be over 55 (35%) and 3.2 million will be over 65 (21.2%). These data indicate that the number of elderly will show both an absolute and relative increase.

Three developments characterize the composition of the category of the elderly: an ample surplus of women, a growing number of unmarried persons and an increasing number of very old persons (80 years and older). The number of very old persons in the category of persons of 55 and older will increase from 11.5% in 1983 to 13.2 % in 2000 and 14.2% in 2030. This is called the process of "double greying".

Between 1984 and 2000 there will be an increase in the number of people of 65 and older and a decrease in the number of people between 0 and 19. The number of people between 20 and 64 years is expected to increase. As a consequence the "demographic burden" ($\frac{0-19 + 65+}{20-64}$) will decline up to the year 2000.

However the question remains, in what respect is the above mentioned ratio adequate. For instance, the costs in the public sector for an aged person are estimated 2.5 to 3 times higher than for a younger person[3].

4.2. the health status of the elderly

In the baseline analysis the health status of the elderly is indicated by (1) life expectancy and life span (2) mortality (by cause of death) (3) morbidity (4) ability to perform Activities of Daily Living (ADL) and (5) subjective health.

(1) In 1982 the life expectancy at birth for men and

women was 72.8 resp. 79.5. The life expectancy at age 65 was in 1981 14.0 for males and 18.5 for females. Only for males is a rise in life expectancy at birth expected to 73.3 up to the year 1990. For females the life expectancy at birth will remain constant.[4]

As far as the life span of human beings is concerned, no important breakthroughs are expected in longevity-technology up to the year 2000. As a consequence the life span will remain 100 to 115 years.

(2) The high infant and childhood mortality rate in the past, usually due to infectuous diseases, has given way to low mortality in the early years of life and a concentration of deaths in middle or late adulthood, typically from chronic diseases (epidemiological transition). As a consequence the survival curve has become more rectangular. The prospects for on-going rectangularization of this curve in the future and related questions of postponement and compression of morbidity, are at this moment topics of international discussion.

The most prevalent causes of death of elderly persons are cardiovascular diseases and cancer.

(3) Empirical data about the morbidity of the elderly in The Netherlands are scattered, because no uniform morbidity registration for the whole country is available. An important aspect of the health status of the elderly is the multi-morbidity (3 to 6 diagnoses per patient). Another aspect is the high prevalence of chronic diseases which require long-term care.

In a recent survey by The Netherland Central Bureau of Statistics (1982)[5], the most prominent disorders mentioned for non-institutionalized elderly of 55 years and older were rheumatism (33%), shortness of breath (21%), hypertension

(20%) and dizziness (15%).

An overall increase in morbidity may be expected in the future as a consequence of the aging of the Dutch population, even at constant age and sex specific incidence and prevalent rates. For instance, general practitioners will be confronted with a rise in morbidity of 5% [6].

(4) In the above mentioned CBS survey [7] the disability to perform the Activities of Daily Living is measured for both institutionalized and non-institutionalized elderly. 4% of the non-institutionalized people need help to perform (some) ADL. For the people in homes for the elderly this percentage is 53%. The last percentage has increased substantially since 1976, when a more stringent admission policy for homes for the elderly came into working. As far as the subjective health is concerned, it must be stated that the subjective health and the objective health are not always congruent. If this is a consequence of "overdiagnosis" on the medical side or "underdiagnosis" on the part of the patient is an open question, especially in those cases where adequate diagnostic criteria are lacking (for example in the cases of hypertension and diabetes mellitus).

## 4.3. health and Social sevices for the elderly

In the baseline analysis the following eight types of health and social services are considered: (1) homes for the elderly (2) nursing homes (3) day care facilities in nursing homes (4) hospitals (5) district nursing (6) home care (7) aged persons' bungalows (8) "district function" of homes for the elderly (i.a. meals on wheels or socio-cultural activities for elderly outside the homes for the elderly).

In addition to the above mentioned professional,

subsidized services, attention will also be paid to informal and commercial services for the elderly. It is expected that the demand for professional subsidized help may be limited by the person's own skills, better housing, and so forth (variables 10 - 15 in figure 1).

The use of services by the elderly has increased a lot in the seventies. Also the overall costs have been growing as a result of three interrelated factors: (1) demographic changes (2) changes in the use of facilities and services and (3) increase in the costs per unit. At the end of the seventies changes in this growth pattern became apparent i.a. as a consequence of government policy. The policy with regard to the (services for the) elderly is directed at enhancing the autonomy of the elderly and delaying institutionalization in nursing homes or homes for the elderly. In this context institutional care has to be restricted and home care and district-nursing have to be expanded (substitution policy). So far the substitution has not been very successful: those formerly cared for by the informal sector benefited from the expansion of home care and district nursing.

We shall not attempt to give an overview of the developments in all eight above mentioned services. As an illustration figure 2 gives an overview of the developments in the number of beds in homes for the elderly in the recent past and the near future.

## Number of Beds in Homes for the Elderly, 1965-2000
## The Netherlands

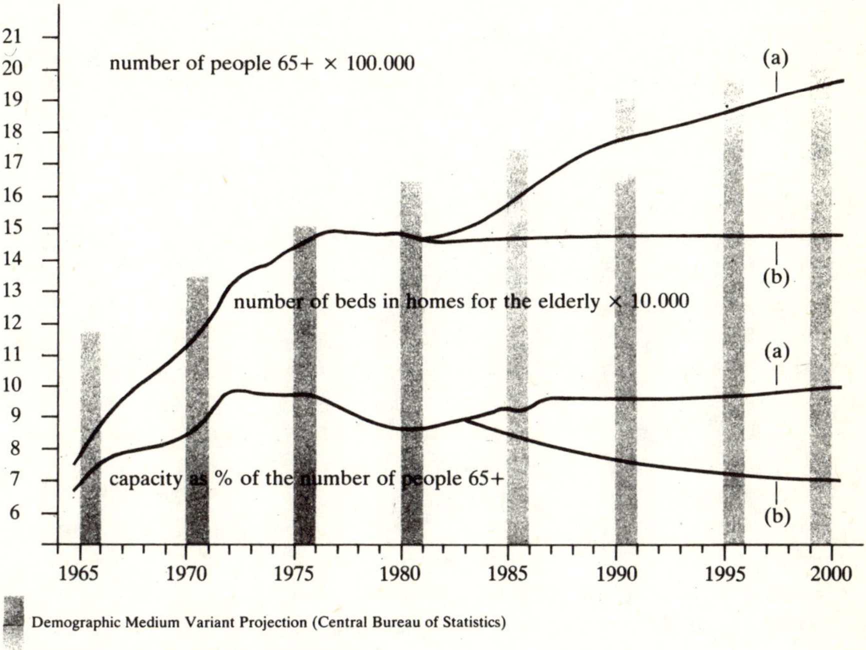


(a) Trend Projections of Level of Care (based on 1979)
(b) Government Policy 1984 (no increase of beds in homes for the Elderly)

*Source: STG Scenario Committee on Aging*

Figure 2

## 5. THREE CONTEXTUAL SCENARIOS

In this paper only a short summary will be given of the contextual scenarios. In the three contextual scenarios the medical, biological and technological developments could be seen to be identical. These developments are treated in Section 6.

### 5.1. Scenario A (Reference Scenario): autonomous developments resulting in about the same pressure on health services for the elderly

In this reference scenario the situation of 1984 is extrapolated towards 2000.

With regard to social developments in this scenario the main processes are:

a) continuation of the emancipation of women (resulting i.a. in a decrease in informal help from women to the elderly);
b) emancipation of the elderly (i.a. as a result of higher levels of education);
c) increasing political influence of the elderly;
d) increase in the proportion of elderly belonging to ethnic minorities;
e) continuation of the debate with regard to, and an increase in, passive euthanasia;
f) decrease in the number of children (important with regard to children helping their parents when they are old and need help);
g) no major changes in norms and values with regard to health and values with regard to health and health services;

In the field of the health services we present the consequences for the demand for health services

a) as a result of purely demographic developments (constant level of care), (for example fig. 2(b)) and
b) as a result of developments in the sex- and age-

specific use of the health sevices in relation to demographic developments.

With regard to the economic aspects of the context scenario we assume that a 0.5% economic growth per year and financial redistribution in the public sector will permit a continuation of the budgets for health services at the level of 1984[8].

5.2. scenario B: autonomous developments resulting in more pressure on health services

In this scenario a number of developments are combined that result in an increase of the pressure on the health services. The B-scenario presents a development in which the demand for health services grows more than could be expected on the basis of the calculations in the reference scenario. In the field of social developments we get (in addition to those mentioned in the reference scenario):

a) growing professionalisation of the care for the elderly;
b) an attitude of the population towards health services which results in a relatively high consumption of these services;
c) less solidarity between the generations (less care by children, less informal help from other informal sources);
d) less solidarity within the generation of the elderly (less informal help from seniors to seniors);
e) norms and values in society put the elderly more and more into an isolated position (disengagement, mandatory retirement);
f) the elderly are less dependent than in 1984, if they are in a better financial position; they can pay for professional help and for commercial help;
g) no new types of living together of the elderly.

The economic context is characterized by a

growth of the national income of The Netherlands by more than 1% a year, as a result of a revitalization of the world economy. As a result of this, the incomes of private households increase, including those of the elderly. The expectations with regard to the health services show a further rise, and the pressure on health services increases.

An increase in the pressure on health services could also coincide, however, with a decline in national income as a result of a worse position of The Netherlands in the world economy, either because of a prolonged general crisis or because of a negative development with regard to the relative position of The Netherlands. If economic stringencies get more tight, about 10% budget cuts could still be met by a more efficient organization of the health services themselves.

The health situation of the elderly would be affected by the increasing pressure on the health services. Professional and commercial assistance may improve, especially for the well-to-do elderly. Self reliance is weak, however, and social isolation is predominant.

### 5.3. scenario C: autonomous developments resulting in less pressure on health services for the elderly

This scenario is characterized by a decrease in the demand for health services as a result of a change in attitudes towards health and related services in the Western world. As a result of demographic developments the absolute demand for health services may still increase, but in any case this increase is less than in the former two scenarios.

The social context shows:

a) a more critical attitude towards health services, diminishing the quest for medical consumption;
b) growth of a sense of responsibility of individuals for their own health, resulting in changes in

lifestyles and less high-risk behaviour; this in turn leads to less cardiovascular diseases, for instance;

c) less pressure on institutional cure and care, the elderly live on their own for a longer time;
d) less solidarity between generations, the traditional care between generations will not return;
e) more solidarity between generation of the elderly; voluntary help from seniors to seniors increases;
f) norms and values in society more and more incorporate the elderly into the dynamics of modern social life;
g) the elderly are less dependent than in 1984: not in the first place because of a better financial position but primarily because the social climate has changed;
h) new types of living together by the elderly are gaining in importance;

The economic context going with scenario C is a slow increase in economic activities on a world scale, resulting for The Netherlands in more than 1% growth of the national income. But again, just as in the case of scenario B, the developments described could coincide with economic decline also. Pressure on health services is related more to the social climate than to the economic activities in the country.

## 5.4. critical incidents

The scenarios A, B, and C together do not provide us yet with a setting for testing strategies that is sufficiently realistic. To enhance the pratical quality of the "testing environment" the scenarios will be combined with "critical incidents". We shall look at a potential postponement of dementia (Section 7). Of course other incidents could be added: a decrease in the number of heart failures or in the number of cancer cases. The second critical incident is a decline in human solidarity (Section 8).

The reader is invited to add his or her own critical incidents to the analysis.

## 6. DEVELOPMENTS IN THE MEDICAL, BIOLOGICAL AND TECHNOLOGICAL FIELD

Information on developments to be expected in the medical, biological and technological field has been gathered in our project by organizing group discussions amongst experts in specific fields, by taking into account the expertise present in the scenario commission on aging and the elderly and by analyzing documents. The information obtained is relatively homogenous. This induces us to present not more than one "context" of autonomous developments with regard to medical, biological and technological developments related to the elderly.

### 6.1. future changes in the health status of the elderly

Certain areas, such as cardiology, urology, pneumology and to a certain extent oncology do not show promising new hypotheses at this moment. There is reason to believe that these areas will not undergo fundamental transformation in the next 20 years [9]. A number of the major diseases afflicting the elderly have a long period of incubation. This applies especially to cancers, cardiovascular diseases, dementia and osteoporosis. Because of the long period of incubation the age cohorts that will become 55 and older between 1984 and 2000 have already acquired (developed) them. In other words it is too late for preventive measures. Only an early diagnosis can help to start treatment at a relatively early moment in order to arrest or to bring about a delay in the disease process. As a result of these developments we conclude that the number of elderly with one or more of these diseases will only increase in the next twenty years due to the demographic pressure (double greying).

### 6.2. developments in biological research

Knowledge of the biological aging process is expected to increase rapidly. This enables us to distinguish natural aging from pathological aging. In the process of aging immunology plays an important role. Immunology may enhance our knowledge of changes in the defence mechanisms of the human body, for instance by elucidating the function of information between cells.

Application of the results of this kind of research is to be expected mainly with regard to diseases with a long incubation period. For this reason the impact on the health status of the elderly can only be modest. Another important field of research is DNA-research. Knowledge of the human DNA-structure enables us to predict risks of acquiring diseases. In the nearer future the DNA-structure can be repaired, influenced (i.a. hereditary diseases) or changed.

### 6.3. developments in medical technology

The applications and possibilities for new medical technology are far from being exhausted. The introduction of microprocessor, laser, new scanning techniques, self-regulatory prothesis (sphincter) will enable improvement in diagnosis, therapy and the quality of life of the patient involved. A problem is the complexity and expense of the new generation of medical equipment.

A second important problem is the multi-morbidity of some of the elderly. This implies that the condition of the older person does not allow medical treatment, despite the fact that age limits for medical treatment of the elderly will continue to go up in the future.

We finish this chapter with a spectacular example of new medical technology. The human body contains more organs with a sensory capability than we are normally aware of. The skin and stomach can sense the difference between dark and light, for instance. Research and development are directed at this time at replacing (lost) human eyesight by a combination of micro cameras as instruments to register information and the stomach to decode it.

## 7. THE POTENTIAL IMPACT OF A POSTPONEMENT OF THE CHRONIC BRAIN SYNDROME

As one of the potential critical incidents we have chosen a postponement of the chronic (organic) brain syndrome (CBS), also called dementia. What arguments have led to this choice?

One set of arguments is related to the relatively heavy impact of CBS on the health status of the elderly and on the health services for this category of the population.

a) CBS is more or less specific for the category of the elderly;
b) incidence and prevalence increase with age; because of the increasing proportion of the elderly in the total population CBS is often called "the epidemic of our century or "the silent epidemic";
c) CBS is an extremely degrading disease;
d) CBS demands a lot of institutional care (10% of the elderly live in an institutional setting; 25% of the demented elderly live in institutions);
e) the life expectancy of the elderly is 75% of the life expectancy of the elderly in general.

A second set of arguments is related to the feasibility of postponing CBS by a number of years in a substantial number of cases:

a) it seems feasible to reach an improvement in the early detection of CBS, by

- better training of general practitioners with regard to this disease;
- a better informed general public;
- better technological facilities (PET and NMR-scan);
- more multi-disciplinary cooperation during detection and early treatment.

b) improvement in the therapeutic field is expected as a result of

- research into the impact of neuropeptides on disorders of the human memory functions;
- research into the impact of transmitter substitution;
- research into non-pharmacological therapies, the socio-therapies ("Reality Orientation Therapy" (ROT), "enriched environment", "structural stimulation";)
- research into the relationship between diet and CBS;
- research into the transplantation of embryonic brain tissue.

An improvement in the etiology of CBS is considered highly unlikely by a large number of experts in the field.

Taking into consideration that improvements in early detection and in early treatment of the first signs of CBS might result in a postponement of (severe) CBS, we decided to analyse the potential impact of a postponement of the CBS by five years, starting in 1990. We used five-year cohorts.

This analysis was hampered by the fact that no Dutch epidemiological empirical data on CBS are available, except for the elderly in institutions. We had to rely on data from other countries with regard to incidence and prevalence.

Applicated to the Dutch situation we estimated that 44.000 men (6.4%) and 73.000 (7.8%) are suffering from dementia (excl. mild). About 12.000 demented persons are living in a nursing home in the early eighties.

The analysis was very complicated. We had to make a restriction to the sector of the nursing homes. In the year 2000 the postponement of the CBS by five years would lead to a reduction of the number of demented patients in nursing homes with 25%. In absolute numbers there will be an increase of the number of patients with 1100 compared with the present number (as a result of the demographic development). Without postponement the number of demented patients in nursing homes will increase with 48% in the period 1983-2000 to 17.700.

The value of this analysis is limited because of the several assumptions we had to make.

## 8. THE POTENTIAL IMPACT OF A DECLINE IN HUMAN SOLIDARITY

A second potential critical incident in our analysis is a serous (further) decline in human solidarity. What will happen if children refuse to care for their elderly parents when they need help ?

Our arguments for choosing this critical incident are:

a) in countries in Western Europe like The Netherlands, there is a growing tendency for individuals to isolate themselves (in their nuclear family setting; in LAT-relationships - Living Apart Together of the young);
b) there is a decline in the number of children per family, and more preference for having no children at all;
c) the emancipation of women is gaining in impetus; as a result of this, daughters (natural or in law) are less inclined to look after their parents (natural or in law);
d) increased geographical mobility of the children.

We analyzed the impact on non-institutionalized elderly of the loss of help from children living on their own, taking 1976 as a baseline. Because of lack of empirical data we had to disregard help from children living together with their parents.

In 1976 [10] 10.3% of the non-institutionalized elderly received help from children who were living on their own. 6.9% received 1 - 6 hours of help each week, 1.9 % received more than 6 hours of help each week, and with regard to 1.5% the number of hours of help per week is unknown.

On the basis of a number of assumptions with regard to the substitution of the help of the children we estimated the increase in the demand for professional help as follows: (situation in 2000)

a) primary care: an increase of 183.000 in the number of elderly of 55 and older living on their own (alone or together with a spouse or partner);
b) homes for the elderly: an increase of 36.000 in the number of elderly of 55 and older living on their own;
c) commercial assistance: an increase of 136.000 in the number of elderly of 55 and older living on their own.

Of course a different set of assumptions would lead to different estimates.

The trends towards less solidarity do not exist in isolation. Some other processes represent a (weak) counter-movement:

a) increase in the flexibility of relationships between the elderly (groups of elderly that live together, informal social networks, LAT-relationships of elderly persons);
b) allocation of compensation to children who are willing to care for their parents when they are old and need help;
c) increase in the (organized) voluntary services as a result of unemployment and shorter working hours.

Of course the decrease in human solidarity with the elderly raises the question as to the extent to which processes of planned change can contribute to an increase in solidarity with the elderly. The social technology with regard to planned change has been improved in the last decades. This improvement is primarily the result of evaluation research into projects of planned change that took place in the not too distant past. These evaluations have shown that successful planned change of human attitudes and human behaviour demands relatively long periods of research, experimental innovation and innovation on a large scale. As a rule successful planned change has required relatively heavy financial support. If too short a period is taken, and if budgets are too small, the ultimate result of the project may be nihil or even negative. Changing lifestyles for instance demands carefully designed long term projects, preferably taking spontaneous processes as a starting point for experimental innovation. Changing patterns of drug abuse or excessive drinking of alcohol may be possible if self-help groups already exist and if these self-help groups can be stimulated. Changing lifestyles in a direction contrary to spontaneous developments, or without some positive spontaneous developments, or without some positive spontaneous processes to start with, is very difficult and has led to a lot of disappointment in the past.

## 9. THREE POTENTIAL PATTERNS OF HEALTH SERVICES

We now come to the second part of the question that has to be answered in our project. It reads: "What are the most important patterns of health services between 1984 and 2000, keeping the future health status of the elderly and their relative number in the population in mind". These patterns of health services are less "autonomous" than the processes dealt with in the previous sections of this paper. The individuals and organizations in The Netherlands responsible for the future of the elderly and especially their health, can shape these health services to quite an extent. In this paper these patterns of health services will be described in terms of "target states". Other aspects of strategy formation, the ways and means to make a choice between these target states and to make the chosen pattern a reality, will be dealt with i.a. in the Policy Document "Public Health 2000".

9.1. business as usual

One opinion would be to continue the health policy that has been followed in the recent past. This health policy has been changed drastically between about 1980 and 1984: expenditures have been reduced as far as possible, for instance. In this potential pattern "business as usual" means continuation of the health policy of 1984, including the changes of recent years.

This pattern is characterized by:

a) slow expansion of health services for the elderly, but no drastic changes in the pattern itself;
b) unmet demands for care and cure; queues are growing, at least in a number of regions;
c) slow expansion of the costs for health services on a national scale, notwithstanding policy measures to curb the rise in costs;
d) decreasing social integration of the elderly in society; no large-scale activities to change this tendency.

9.2. first class cure and care

A second option would be an emphasis on improving the quality of cure and care, and on expanding the health services. This option would require a larger share of the national income for health services than was allocated by parliament in 1984.

The characteristics of this pattern in short:

a) rapid expansion of health services for the elderly,
b) decrease in the self reliance of the elderly,
c) avoidance of queues,
d) rapid expansion of the costs for health services for the elderly on a national scale; but: a larger number of operations could result in a decrease of costs per operation, i.a. because of economies of scale,

e) decreasing social integration of the elderly in society.

The "philosophy" behind this pattern of health services is sometimes characterized as "medicalisation of society". Care and cure invade more and more aspects of modern life.

9.3. towards a selfcaring society

The third package of health services would imply a change, compared with health policies of the early eighties. Now a moderate improvement of the health services themselves would be combined with a major effort to change the social conditions of the elderly in society. Improvements not only within the medical system itself, but also in the surrounding social networks.

In short the charateristics of this pattern are:

a) systematic attempts to concentrate health services for the elderly on meeting "vital" medical needs;
b) systematically attempting to make the elderly more and more self-reliant, in a medical and in a social sense;
c) systematically attempting to activate human solidarity within the category of the elderly (helping each other) and within society (redistribution in favour of the elderly);
d) systematically attempting to change life styles of the elderly themselves and of the cohorts entering "third life";
e) systematically attempting to keep costs within (politically) acceptable limits.

The option implied by "towards a selfcaring society" would stand a chance, if the "state of the art" or planned social change was taken seriously, and if major effort was made on the basis of the knowledge and know how available to induce social change.

## 10. CROSS-IMPACT OF AUTONOMOUS DEVELOPMENTS AND PATTERNS OF HEALTH SERVICES

In this section the cross-impact of the autonomous developments and the potential patterns of health services will be analyzed. This will be done not by a traditional cross-impact analysis, but by a more qualitative approach. On a number of aspects a quantitative analysis is possible, for instance with regard to the critical incidents described in foregoing sections.

### 10.1 impact of contextual scenario A

A continuation of the pressure on health services compared with 1984 would imply in short:

a) pattern 1 would lead to serious financial and social difficulties; one, or a small number, of major medical innovations would reduce these difficulties somewhat but not drastically; further decrease of informal aid to the elderly would make the situation even more difficult;
b) pattern 2 would lead to the same problems as pattern 1, but in a worse way;
c) pattern 3 would give a chance to cope with the problems resulting from contextual developments A.

### 10.2. impact of contextual scenario B

An increase in the pressure on health services compared with 1984 would lead to, sketched briefly:

a) pattern 1 would imply even more financial and social difficulties than in contextual scenario A,
b) pattern 2 would be simply disastrous,
c) pattern 3 would stand a very small chance only.

10.3. impact of contextual scenario C

a) pattern 1 would stand a small chance,
b) pattern 2 would lead to serious difficulties,
c) pattern 3 would give the best chances in the nine situations included in the analysis.

## 11. THE RESULTS OF THE SCENARIO PROJECT AS A CHALLENGE

The scenarios provide us with a number of prospects.

11.1. the majority of the elderly between 1984 and 2000 will lead active lives, will enjoy good health and will live in their own homes;

11.2. the patterns of autonomous developments that may be expected for the next fifteen years are all relatively unfavourable for the future of the elderly in The Netherlands;

11.3. even a cumulation of breakthroughs in the medical field would not substantially improve the health status of the elderly in general.

11.4. <u>the scenarios as instruments to enhance the awareness of the problems involved</u>

As we saw, improving the future of the elderly has to rely heavily upon public support, political support and support from the health services. The support will have to be mobilized. This mobilization can only succeed if the potential supporters can be made aware of the problems that face the elderly in the years 1984-2000 and beyond.

We are inclined to overestimate the favourable developments, and to underestimate the negative prospects. We are also inclined to consider alternatives that combine favourable developments that will never coincide in reality. To avoid these mistakes scenarios will have to be utilized.

## 11.5. the scenarios as instruments in the future

Our scenario project was medium-sized (1 - 5 man-years). Up to now scenario projects have as a rule been large scale enterprises, requiring five man-years or more to complete. Smaller organisations in the developed countries, and the developing countries, cannot work with large scale scenario projects. In order to broaden the applicability of scenarios, medium-sized and mini-projects will have to be developed. In a medium-sized project, one to five man-year is enough. Mini-projects need less than one man-year to complete. Sometimes even instant projects are possible, requiring one man-week or less.

If monitoring of social developments by social indicators provides an empirical basis for designing scenarios, and if theories are available to explain the processes taking place, medium-sized and small scenario projects are feasible. On the other hand: multi-sectoral scenarios to improve national and international health planning and management will also require large scale scenario projects in the future.

## 11.6. the scenarios as a "learning environment" for developing strategies

Strategies with regard to the future health services for the elderly have to be developed under stress. Time is short. Financial margins are getting more and more narrow. Instruments to change human solidarity and lifestyles in a systematic way are scarce, expensive, time-consuming and still in the stage of development.

This situation requires careful design of strategies, and it requires assessment of potential effects and side-effects of strategies under consideration. The strategies will be developed in the Policy

Memorandum "Public Health 2000". the scenarios can provide a "learning environment" for designing and assessing these strategies. They can also provide a framework for the interpretation of the strategies that will appear in the Policy Memorandum.

In the near future the policy proposals of "Public Health 2000" will have to be put into practice. These processes of planned change can profit from the scenarios. The scenarios can help to achieve flexible planning and management for the health of the elderly.

## NOTES

1) CBS: Prognose van de bevolking van Nederland na 1980. Den Haag 1982

2) CBS: Prognose van de bevolking van Nederland na 1980, Den Haag, 1982

3) De Kam, C.A. en F.P. van Tulder: Wie dan leeft, die dan zorgt? In: Ouderen en Sociale Zekerheid, Den Haag, 1983

4) CBS: op cit.

5) not yet published material

6) Hoogen, H.J.M. van den, E.A.: Huisarts en Vergrijzing. In: Tijdschrift voor Soc. Geneeskunde 60 (1982), nr. 25, p. 870-874

7) CBS: op cit.

8) SCP: Collectieve uitgaven en demografische ontwikkeling, 1970-2030, Den Haag, 1984

9) C. Burg: "The prospects for medicine in the next 20 years" (WHO report, July 1983)

10) CBS: De leefsituatie van de Nederlandse bevolking

van 55 jaar e.o. Deel 4: Wonen en validiteit. Den Haag, 1983

# THE UNPREDICTABILITY OF POPULATION TRENDS

Nico Keilman*

ABSTRACT

Population developments are inherently unpredictable, even for a period of a few years. Therefore, in order to produce the necessary demographic numbers, the notion of a "forecast" is often used. But official population forecasts rely on deterministic models, that cannot quantify uncertainty.
This paper discusses methods to deal with uncertainty in current population forecasting. Sources of uncertainty are identified, and the role of variant forecasts is discussed: alternative futures or uncertainty variants? We treat the issue of growing uncertainty with increasing forecasting horizon: forecast or mere projection? Some ideas are given on the presentation and use of population forecasting results as inputs to planning.

Forecasting practice and forecasting errors in The Netherlands serve as an illustration. Finally, major research topics are identified. Most prominent is the forecast of forecasting errors, in particular the separation between the impact of forecasting methodology and that of current demographic trends upon forecasting accuracy.

KEY WORDS

population forecasts, uncertainty, projection, forecasting practice, forecasting errors

---

* Netherlands Interuniversity Demographic Institute (NIDI)
P.O. Box 955
2270 AZ Voorburg
The Netherlands

ACKNOWLEDGEMENTS

Thanks go to Willemien Kneppelhout for editing the English text of this paper and to Tonny Nieuwstraten for skillfully typing the manuscript.

## 1. WHY DON'T POPULATION FORECASTS COME TRUE?

Demography has always been a social science in which quantification plays a relatively important role. The subdiscipline of analytical demography, or demometrics, was developed long before econometrics, psychometrics and sociometrics. Due to the fact that demography often uses exact reasoning, one might easily get the impression that causal relationships can be found and that reliable predictions can subsequently be made. This is a misconception.

Demography studies individual behaviour and the relationships between individuals. Thus an attempt is made to predict individual and group behaviour. Cf. quantum physics, a science in which the characteristics of minute particles are explained in order to learn something about the matter which is made up of these particles. In 1930, Heisenberg formulated his famous Uncertainty Principle: it is not possible to determine simultaneously and with absolute certainty the position and velocity of a particle. If we measure the exact position, the velocity is distorted, and vice versa (Heisenberg, 1953, p.30). It is therefore impossible to be 100% certain about the actual movement of individual particles, let alone to be able to predict their movement. If we describe the behaviour of such a particle we must bear in mind that chance plays a role. Only a large amount of such particles, such as a given volume of oxygen, can be described with sufficient certainty: at a given aggregate level the element of chance has been reduced to such an extent that it can be neglected. Heisenberg's Principle of Uncertainty undermined the prevailing ideas on mechanisms of chance and causality. Probability laws were no longer considered a product of human ignorance. Laplace's determinism and the belief in perfect predictability (derived from Newtonian astronomy) were abandoned. A certain degree of indeterminism or uncertainty started to become accepted. Thus the deductive-nomological interpretation of causality, explanation and prediction changed into an inductive-statistical approach.

We can say that, analogous to the behaviour of elementary particles, it will never be possible to predict the behaviour of individuals with absolute certainty. Should we wish to describe such behaviour, we will have to resort to probability laws. This does not mean that our ignorance is short-lived: it is an established fact that human behaviour is indeterminate.
Therefore, demographic behaviour cannot be predicted: we cannot give a 100% accurate statement on future demographic trends. Since uncertainty plays a role, we can only make a forecast: a plausible and realistic estimate of the future based on our knowledge of the present. As with weather forecasts, the reliability and durability of such demographic forecasts are limited (Tennekes, 1984, 1984, p.17).

All this means that when making population forecasts, we should continue to try to gain a better understanding of our demographic behaviour but we should also (even more than in the past) study the uncertainty that accompanies population forecasting. This is particularly important in view of the desired optimal use of population forecasts. It is surprising that forecasting uncertainty has so rarely been the subject of research. One of the reasons for this is the relatively long period of time needed for population forecasting (Keyfitz, 1984, p. 12). Some 30 years are necessary for a thorough study of the reliability of such forecasts. Meteorologists only need a week in all to make a forecast, to compare it with the real situation and to make a new forecast; economists need anything from several months to a year; population forecasts may need as long as a generation. A second reason is that demographic forecasts, particularly those at the national level, are often officially authorised. This, added to the fact that a relatively small number of demographers are concerned with the future, means that competing forecasts are lacking. Thirdly, forecast users are not used to working with uncertainty. They want future population statistics (which form the basis of their planning) to be available as a single figure. They do not like working with probability distributions. Finally, many users of population forecasts are specialised in a particular field, such as housing, regional policy, transport, energy, economics etc. For them, uncertainty regarding future population trends is far less important than the uncertainty in their own specific field.

This paper gives an overview of the role of uncertainty in population forecasting. We will discuss the most important causes of forecasting errors, as well as the instruments used to measure uncertainty in deterministic and statistical analyses. Examples are derived from the post-war population forecasts of the Netherlands Central Bureau of Statistics. Although these forecasts apply to the national level, the points discussed here also are valid for regional and categorical forecasts.

## 2. SOURCES OF FORECASTING ERRORS

Population forecasts never come true, there is no doubt about that. On second thoughts, you never know when dealing with chance. It would therefore be better to say: the probability that a forecast will come true is zero. The terms forecasting error and uncertainty are closely connected to each other. We could put it like this: a forecasting error is observed uncertainty; uncertainty refers to the future, whereas a forecasting error refers to the past (Schéele, 1981, p.3).

The process of population forecasting is made up of several steps. It is worthwile trying to find out in which stages of this process forecasting errors are usually generated. This can be done by distinguishing the following five types of errors (Hoem, 1973; Keyfitz, 1977,p.227).

1. Registration-, rounding-off and estimation errors of the observed trends. In countries like The Netherlands, such errors hardly have a bearing on the quality of the population forecast.
2. Randomness in parameters. Stochastic fluctuations

in the estimated numbers of births, deaths and migrants are not taken into account in the current deterministic forecasting methods. However, a number of different studies have shown that errors caused by such simple extrapolations of (supposed) trends are very small (e.g. Sykes, 1969; Schweder, 1971).

3. Erroneous trends in forecasting parameters. If the quantitative course of real demographic trends differs substantially from the expected course, significant forecasting errors may result.
4. Sudden shifts in the parameters. Large forecasting errors can also be the result of extreme circumstances such as wars, disasters, serious economic depressions and the like. Such circumstances are not very common, however.
5. Inaccurate model specification. If one simulates the population trends over a known period with the aid of a forecasting model, the results should ideally correspond to the observations. In the past, the international migration component was not always incorporated in national population forecasts, often resulting in rather serious forecasting errors. Now that international migration is taken into account, the models seem sufficiently valid.

The most profound forecasting errors are therefore caused by erroneous assumptions regarding trends in fertility, mortality and migration. Two stages in the assumption-making process account for this in particular. Firstly, when formulating future social and socio-demographic developments. For example, statements such as "the position of the family in society will become less dominant", or "structural factors such as advancing emancipation and growing individualism contribute to the continuing fertility decline; an increasing number of women are following higher education and wish to practise in their field for as long as they possible can. This trend is likely to continue". Secondly, important errors are generated when formulating the so-called key assumptions, i.e. when social and socio-demographic hypotheses are translated into quantitative demographic indicators such as the average number of children per woman, the average life expectancy etc. Fertility will remain low in the future. But will the average number of children in 1990 be 1.3, or 1.5, or maybe 1.7? We don´t know.

From the above, it is clear that the hypotheses used in a population forecast, in particular general assumptions and key assumptions, have a greater bearing on the quality of the forecast than the accuracy of the model. If the key assumptions are correct, the model is of secondary importance. But if they do not adequately reflect the future, a good model cannot save the forecast (Ascher, 1979, p.199).

## 3. UNCERTAINTY IN DETERMINISTIC ANALYSIS

At present, national population forecasts are made with the aid of deterministic models. Despite the fact that randomness (quantified uncertainty) is lacking in these models, they can give us a deeper - qualitative - insight into the uncertainty which is inherent to statements concerning the future. Three instruments are used for this purpose: variants, border-years and sensitivity analyses. The presentation can also express the degree of uncertainty. These four aspects will be discussed below.

### 3.1. Variants

As early as 1933, Thompson an Whelpton published several variants of their estimates of the future population growth in the United States because they were not sure about the course the demographic components would take. Ever since, this has become common practice, stimulated, amongst others, by the United Nations. In the 1950´s and 1960´s, the Netherlands

Central Bureau of Statistics published one or two experimental variants, alongside the main variant in its population forecasts. It was clear however - by the presentaion of the results - that these secondary variants were considered less important than the main variant. Not until the 1970 forecast, did all variants carry equal weight.

Two points should be taken into consideration when choosing the components for which more than one hypothesis should be made (CBS, 1982, p.17).

1. The significance of each individual component for the results of the forecast. A 10% variation in the average number of children per woman generally results in greater fluctuations in the forecasting results than a 10% variation in the probability of on eventual divorce.
2. The past course of the components. The more erratic the observed development of a given phenomenon, the more difficult it is to estimate its future course.

The first consideration may be used when the outcome of the sensitivity analysis (to be discussed at a later stage) is known. The second consideration has to do with the fact that we know little about the quantitative relations between the demographic components and the underlying social, economic, cultural, political and technological factors. Here, we are interested in quantitative answers to questions such as:

- how does the increased emancipation of women influence the number of children a woman will bear?
- in which way does technological development influence mortality?
- to which extent will the demographic behaviour of cultural minorities presently living in our country differ from that of the total population in the future?

If the quantitative relations referred to here are not sufficiently known, this does not mean that the social and socio-demographic trends which influence the forecasting components should not be taken into account. In such cases assumptions are indeed made, even though they are merely qualitative. However, due to the fact that quantitative insight is lacking into the relationships between social and socio-demographic trends on the one hand and the course of the demographic indicators on the other hand, no more than an extrapolation of the indicators can be made; of course the qualitative course of the indicators computed earlier is taken into account, Therefore, as long as a causal model for demographic components is not available, (in which - if ever found - chance is likely to continue to play an important role) we can, as a second-best-approach, simply look for stability within these demographic components.

In view of the above considerations, the CBS has, during the past years, used several quantitative key assumptions for birth, nuptiality, and international migration. For each of these components a high, a medium and a low assumption have been drawn up. By combining the 3 possibilities for the 3 components, 27 forecasting variants could, in theory, be computed. In practice, only the two extreme variants and the medium variant are used (CBS, 1982, p.17). The medium variant may be interpreted as the most plausible alternative at the time the forecast is made. Of course, the high and the low variants should also be plausible. The three variants are related, as is shown by the fact that they are all derived from only one set of general hypotheses and qualitative key assumptions. Such variants can be called "uncertainty variants". Their raison d´être is the uncertainty that accompanies the act of quantifying qualitative key assumptions.

When several qualitative key assumptions and general hypotheses are used, we speak of "alternative futures". Up to a certain extent, this was the case in the 1975 CBS-forecast, in which two variants were used for fertility (see table 1).

Table 1. Average number of children per woman, assumptions used for the 1975 CBS-forecast

| | Marriage cohort | | |
|---|---|---|---|
| | 1955 | 1970 | 1980 |
| Low variant | 2,5 | 1,7 | 1,4 |
| High variant | 2,5 | 1,8 | 1,9 |

Source: CBS (1976, blz. 33)

The low variant assumed a continued fertility decline in The Netherlands. The high variant, on the other hand, assumed a slight recovery following an initial decline. Consequently, the high variant spoke of postponement and catching-up of births, whereas the low variant spoke of a (partial) cessation.
It speaks for itself that any interpretation that variants are ´alternative futures´ and our definition of a forecast as being the most realistic future population trend, are contradictory.

The distance between corresponding variables in the high and in the low variant is indicative of the degree of uncertainty of the phenomenong in question. For example, the distance between the variants for international migration, known to be a rather erratic phenomenon, is quite large. In the 1984 forecast, the CBS assumed a net migration surplus for 1985 with a margin of 116% around the level of the middle variant. By 1990 the margin has reached 600%. The margin for the fertility is much smaller: 27% for the 1985 cohort. When determining the distance between the high and the low cohort, one must bear in mind that a wide margin is more reliable, but that it does not give precice information.
"In 1986 the TFR will fluctuate between 1 and 2" is a reliable statement, but hardly precise. For the user, such ´it-may-happen-or-it-may-not´ type of statements are not very valuable.

Table 2. Net migration surplus and average number of children per woman, assumptions of the 1984 CBS-forecast

| | Year | Low | Medium | High | Difference 1) |
|---|---|---|---|---|---|
| | | | | | % |
| Net migration surplus as a result of international migration | 1985 | 2798 | 6666 | 10.534 | 116 |
| | 1990 | 5000 | 2500 | 10.000 | 600 |
| Average number of children per woman, birth cohort 1985 | | 1,3 | 1,5 | 1,7 | 27 |

1) High-low difference as a percentage of medium
Source: CBS (1985)

Therefore, one of the most difficult tasks facing the forecaster is to arrive at an acceptable compromise between precision - the distance between the variants - and reliability - the probability that a forecast comes true (Van Dantzig, 1952, p. 197). A good balance between the two can only be found, in fact, if we know how the forecast will be used. Since national forecasts are always "general-purpose" calculations, the forecasters will have to find an intuitive answer to the precision-reliability dilemma.

## 3.2. Border-years

The more distant the future, the greater the uncertainty. In population forecasts, uncertainty, which grows with time (sometimes very rapidly) can be attenuated in two different ways. Firstly, by enlarging the difference between the high and the low

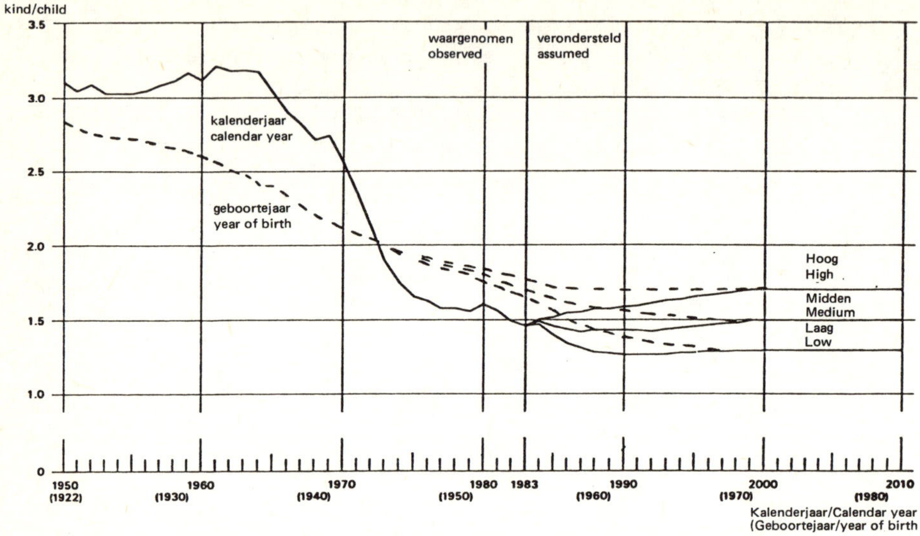


Figure 1. Average number of children per woman by calendar year and year of birth respectively

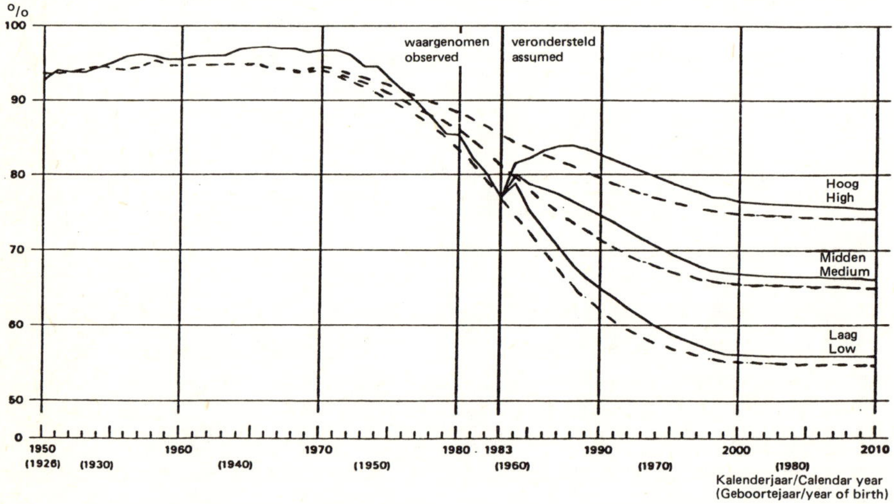


Figure 2. Probability of first marriage for females by calendar year and year of birth respectively

Source: Cruijsen (1985), p. 33, p. 35.

variant. Secondly, by defining a so-called border-year for each component. We will here deal with the latter in more detail.
In the 1984 forecasts of the CBS, the future development of a component such as mortality is formulated until the year 1995. After this year the hypotheses remain constant, since it is barely possible to select a plausible mortality trend from a large range of possible future trends.

The greater the uncertainty of a component, the closer the border-year will be to the year in which the forecast is made. The border-year chosen for international migration in the 1984 CBS-forecast was 1990. For fertility, nuptiality and remarriage the year chosen was 1995. Marriage of never-married persons and fertility remain constant after the year 2000, partly as a result of the fact that the course of these two phenomena, when analysed longitudinally, is fairly regular (see figures 1 and 2).

After the year 1990, in particular after the turn of the century, estimations are no longer called ´forecasts´ but are gradually known as "experimental calculations".

3.3. Sensitivity analyses

It is not always clear whether or not the selected course of an input variable is plausible. If a given variable does not influence, or hardly influences the forecast results there is no reason for concern. If, however, a 10% change in the variable drastically influences the results, several variants ought to be given. One can distinguish between important and unimportant variables by applying a sensitivity analysis. Such an analysis examines the effect of a 1% change in the input variables on the results of the forecast. This can be done analytically or empirically. In the analytical method, the formulas derived from differential calculus are applied to the model´s formulas. The analytical approach usually looks at the short-term effect of a one-off change of a single input variable on the forecast results. When long-term trends are to be examined, the analytical approach is very complicated. In such cases computer simulation is used. This method of calculating sensitivities is called the empirical approach.

Cruijsen and Van Hoorn (1983) carried out a sensitivity analysis for the 1980 CBS-forecast. They examined disruptions in the total population size caused by fluctuations of the input variables. According to these authors, the most crucial variables are:

1. marital fertility rates of parities 1 and 2 for females aged 20 to 30 years;
2. marriage rates for never-married males persons of approximately 22 years;
3. net international migration rates for persons between 0 and 30 years of age;
4. mortality rates for males over 50 years and for females over 55 years.

Together, these input variables account for about 15% of the total number of variables. Extra attention should be given to these variables, both when making the hypotheses and while the forecast is being monitored.

We have now spoken about the analysis of the influence of changes in the input variables on the forecasts. Such a sensitivity analysis examines the assumptions. The impact of changes in the model structure can also be studied. Such a sensitivity analysis is almost always empirical.

Nelissen and Vossen (1983) have recently designed a short-term population forecasting model based on regression results where the demographic components were the dependent variables and the independent variables were the socio-economic, cultural and socio-psychological phenomena. Although the methodology of these authors is entirely different from that applied by the CBS, both approaches arrived at a rise in fertility at the end of the period 1980-1984 (see table 3)

Table 3. Total fertility rate 1980-1984, estimations of Nelissen and Vossen, of the CBS, and observed

| | 1980 | 1981 | 1982 | 1983 | 1984 |
|---|---|---|---|---|---|
| Nelissen and Vossen (1983, blz. 76) | 1,530 | 1,499 | 1,462 | 1,475 | 1,545 |
| CBS (1984, blz.44) - Medium Variant | 1,601 | 1,582 | 1,617 | 1,627 | 1,696 |
| Observed | 1,600 | 1,559 | 1,496 | 1,470 | 1,49*) |

*) Provisional figure

By comparing the forecasts with the observed figures, one can say that both forecasts regarding fertility were too high. We cannot, as yet, determine whether Nelissen and Vossen and the CBS have (prematurely) predicted a recovery which might still take place in the future.

It is surprising that there are so few competing forecasting models in Dutch demographic practice, as opposed to Dutch economic practice. It is advisable to follow the course taken by Nelissen and Vossen, paying particular attention to two points:

1. what causes the differences between the forecast results found by the various models?
2. how reliable are the forecasts of these models?

## 3.4. Presentation of the forecast

By presenting the forecasts in the correct manner the forecast-user can be made aware of the uncertainty of the results. The user should realise that he must not apply a planning methodology which assumes that population forecasts are very precise. It is of the utmost importance that decisions which have been based on population forecasts are robust against possible forecasting mistakes, or else they should be sufficiently flexible so that they can be adjusted to revised forecasts (Baxter and Williams, 1978, p.68).

If more than one variant is used, each one should be given the same amount of attention. Incorporating secondary variants in an appendix therefore serves little purpose. The results of uncertainty variants ought, where possible, to be mentioned simultaneously. Therefore, a statement such as "In the year 2000 our country will number 14.8 to 15.5 million people" is much more effective than "According to the medium variant our country will number some 15.1 million people in the year 2000; the low variant puts the number at 14.8 million and the high variant at 15.5 million".

This uncertainty can be emphasized in tables by placing the variants side by side, as shown in table 4. By doing so, the user is more strongly confronted with the uncertainty than would be the case if three separate tables had been constructed. In elaborate tables, such as age-structure tables, this is not always possible since the table will become too confusing and therefore less useful.

Finally, presenting information on the size of a certain population category in units of a thousand is usually sufficiently precise; for information regarding the distant future, however, units of ten thousand or a hundred-thousand are to be preferred. An analysis of the degree of reliability for the population groups and the forecasting periods distinguished, can suggest which unit ought to be used. In view of the rapidly decreasing reliability of forecasts for periods exceeding 15 years, say, one should excercise the greatest care when publishing long-term forecasts.

Placing the results of a forecast in a historical context, can also give us an impression of the uncertainty. The more (often) one has to adjust the forecasts, the greater the uncertainty. The following passage, taken from 1984 CBS forecast (Cruijsen, 1985, p.40) illustrates the above: "Only three years ago the population of the Netherlands in the year

Table 4. Population size and population growth in The Netherlands, 1984 CBS forecast

| | Bevolking op 1 januari | | | Bevolkingsgroei[1] | | | | | |
|---|---|---|---|---|---|---|---|---|---|
| | | | | absoluut | | | relatief | | |
| | L | M | H | L | M | H | L | M | H |
| | x 1 000 | | | | | | % | | |
| 1984 | 14 395 | 14 395 | 14 395 | 57 | 60 | 62 | 0,40 | 0,41 | 0,43 |
| 1985 | 14 452 | 14 454 | 14 457 | 47 | 58 | 68 | 0,33 | 0,40 | 0,47 |
| 1986 | 14 499 | 14 512 | 14 525 | 38 | 55 | 73 | 0,26 | 0,38 | 0,50 |
| 1987 | 14 537 | 14 567 | 14 597 | 33 | 54 | 75 | 0,23 | 0,37 | 0,51 |
| 1988 | 14 570 | 14 621 | 14 672 | 29 | 52 | 76 | 0,20 | 0,36 | 0,52 |
| 1989 | 14 599 | 14 673 | 14 748 | 26 | 51 | 77 | 0,18 | 0,35 | 0,52 |
| 1990 | 14 624 | 14 725 | 14 825 | 23 | 50 | 77 | 0,16 | 0,34 | 0,52 |
| 1995 | 14 729 | 14 968 | 15 209 | 16 | 44 | 72 | 0,11 | 0,29 | 0,48 |
| 2000 | 14 771 | 15 147 | 15 526 | −5 | 20 | 46 | −0,04 | 0,13 | 0,30 |
| 2010 | 14 466 | 15 075 | 15 692 | −56 | −33 | −10 | −0,39 | −0,22 | −0,06 |
| 2025 | 13 260 | 14 293 | 15 377 | −107 | −74 | −37 | −0,81 | −0,52 | −0,24 |

[1] Voor het jaar vermeld in de eerste kolom van de desbetreffende regel.

2030 was estimated at 14.3 to 16.7 million. This forecast has meanwhile been put at 12.7 to 15.2 million. This example adequately shows that the value of long-term forecasts is limited".

## 4. UNCERTAINTY IN STATISTICAL ANALYSIS

Although classical forecasting techniques acknowledge that uncertainty exists, they do not show they do not show what the degree of uncertainty is. In other words, is the probability that the population size will amount to 14.8 to 15.5 million 30%, 60% or 90% in the year 2000? This cannot be determined without statistical techniques.
The uncertainty can be quantified in different ways. The three most important methods are time series models, Markov models and the analysis of forecasting errors.

### 4.1 Time series models

A large number of authors have tried to analyse demographic time series, in particular for fertility, with the aid of Box-Jenkins-type methods (Saboia, 1974, 1977; Lee, 1974; McDonald, 1979, 1981; De Beer, 1984; Brunborg, 1984). The internal structure of such a time series can usually be adequately modelled. The resulting forecasts are sometimes even better than those found with the aid of traditional techniques. Such an approach has a number of important drawbacks however.
First of all, a time series model for fertility is only suitable for short term forecasts. The confidence interval increases sharply: after 5 years the 95% interval constitutes some 30% of the point estimate, after 15 years roughly 50%. In addition, such a model is based on stationary fertility trends, which in reality only exist for short periods of time. Secondly, time series models may yield alternative models which differ considerably, yet it is often

impossible to make an a priori choice between them. One can therefore conclude that time series models quantify the uncertainty satisfactorily, but that they should be used with care. They are particularly useful for determining short-term confidence intervals, i.e., the distance between the high and the low variant for the first years of a forecast.

## 4.2. Markov models

The second approach to quantifying uncertainty in population forecasts is by way of Markov models and their generalisations. The variables of the forecasting model are, in fact, made stochastic. Either the Leslie matrix (including fertility and survival rates) is written as a stochastic matrix, or else a vector with random fluctuations is added to the population vector. In both cases the forecast - made according to the classical approach - can be seen as the average of the stochastic forecast (Feichtinger, 1971, p. 8). The studies of Pollard (1966), Sykes (1969), Schweder (1971), Le Bras (1974) and Schéele (1981) are based on this principle.

The conclusions presented in these studies are not very encouraging. If we restrict ourselves to random fluctuations of the fertility and mortality rates (the second source of forecasting errors - see section 2) the effect is minimal. The biggest mistakes are made because of erroneous assumptions regarding the future trends of the input variables. When such a trend is described in a stochastic process the mathematical problems become formidable: it is no longer possible to derive simultaneous confidence intervals analytically. Therefore, elaborate simulations will be needed.

Valid modelling of the components will remain problematic, however. The classical approach seems more effective for estimating the trend - in particular in the long term - than an approach that uses stochastic processes. The advantages of both approaches can be combined by applying so-called subjective

probability distributions. As is the case in traditional forecasting, the future course of a number of characteristic input variables of each component is determined subjectively; for example, the average number of children and the age distribution at birth. These variables are considered to be parameters of probability distributions. The form of this probability distribution has been chosen a priori. In a number of simulations a whole range of forecast results may be produced by letting the computer draw from the given probability distribution at each simulation.
Pflaumer (1984) constructed a stochastic Leslie-matrix for West Germany with the aid of subjective probability distributions. He assumed that the TFR would be at least 1.2 and at the most 1.8, with a median of 1.4 children per woman. He presupposed that in two consecutive five-yearly periods the realisations of the TFR did not differ by more than 0.5 children per woman; the median of the difference was 0.3 children per woman. Thus an autocorrelation structure was added to the TFR. Pflaumer made analogous assumptions for mortality and international migration. He determined, by way of a Monte Carlo-simulation, that in the year 2050 the probability that the population of West Germany will lie between 35.4 and 44.3 million people is 90%. In the year 2000 this 90% interval will amount to 57.2 - 60.7 million.

### 4.3. Analysis of forecast errors

Researchers working with time series analyses and Markov models were primarily concerned with explaining observed trends. However, if one is interested in forecast uncertainty, existing forecasts should be evaluated and the forecast errors found should be analysed. If, by doing so, a certain degree of regularity is found, a forecast can be made of these forecast errors.

Within the field of demography, Keyfitz (1981) and Stoto (1983) have elaborated on this idea. They analysed the errors in the average annual growth rates for the total population found in a large number of forecasts for a number of different countries. From this, the expected standard deviation for future forecasts was calculated. If we apply this method to the post-World War II CBS forecasts. and compare the results with the observed population size over 1950-1984, we get the following results (Keilman, 1983).

Table 5. Forecast errors of The Netherlands' population 1950-1984, average annual growth rate (in percentage points)

| | |
|---|---|
| Average error, including sign | -0.02 |
| Average absolute error | 0.18 |
| Standard deviation of the error | 0.22 |

The average error shows that on the whole the forecasts in the period 1950-1984 only deviated slightly from the observed total population. The average is negative, which means that the actual numbers have been underestimated. The average absolute error is much bigger than the average error. Therefore, the underestimations in the period 1950-1965 have amply compensated for the post-1965 overestimations. The 0.22 percentage points of the standard deviation compares favourably with the 0.29 percentage points found by Keyfitz for countries resembling the Netherlands. It was apparently relatively easy to make a forecast in our country.

If we wish to say anything about errors in future forecasts we had best restrict ourselves to our experiences of the 1970's. As a result of the abrupt fertility decline after 1964, the forecast errors were excessively large in the 1960's. In the 1950's on the other hand the forecast errors were relatively small because the demographic trends were very regular.

In the 1970's the average forecast error was 0.09 percentage points; the population increase was

overestimated. The standard deviation of the error was approximately 0.15 percentage points. If we assume that the 1984 forecast will be neither worse nor better than the average post-1970 forecast, then the probability that the population will amount to anything between 14.7 and 15.5 million in the year 2000 will be about 2 to 1 (66%).
In figure 3 this 2/3 confidence interval is compared with the variants of the 1984 CBS-forecast. The odds that the CBS-forecast comes true are at an almost constant level of 2 to 1. A similar value has recently been found for the population forecast of the United States (Land, 1985,p.11).

The evaluation of forecast errors ought to be much more refined than the procedure described above. Firstly, the forecast results used should include variables other than just the total population size. Secondly, the analysis of forecast errors should be more thorough than in the past. Several standards should be used to evaluate the difference between a forecast and an observation. Granger and Newbold (1977, p. 278-289) have discussed how macro-economic forecasts ought to be evaluated. Kuijsten (1984) studies the suitability of a standard introduced by Keyfitz, namely the "quality of prediction" (sic). In addition, a time series analysis of the forecast errors could lead to better forecasts of these errors, than the forecasts made according to the method of Keyfitz and Stoto.

Decomposing the forecast errors for the total population by age group, yields the following picture for the post-war CBS-forecasts (see figure 4).

We see an overestimation of the young age categories and an underestimation of the older age categories. There is a clear overestimation of fertility in the 1965 to 1972 forecasts. The underestimation of the aged population is the result of excessively pessimistic assumptions regarding mortality. As the forecast period rises, forecast errors grow, but for ages between 10 and 74 years the error does not exceed plusminus 2%. The forecast error grows quickly for the older age groups in particular where the

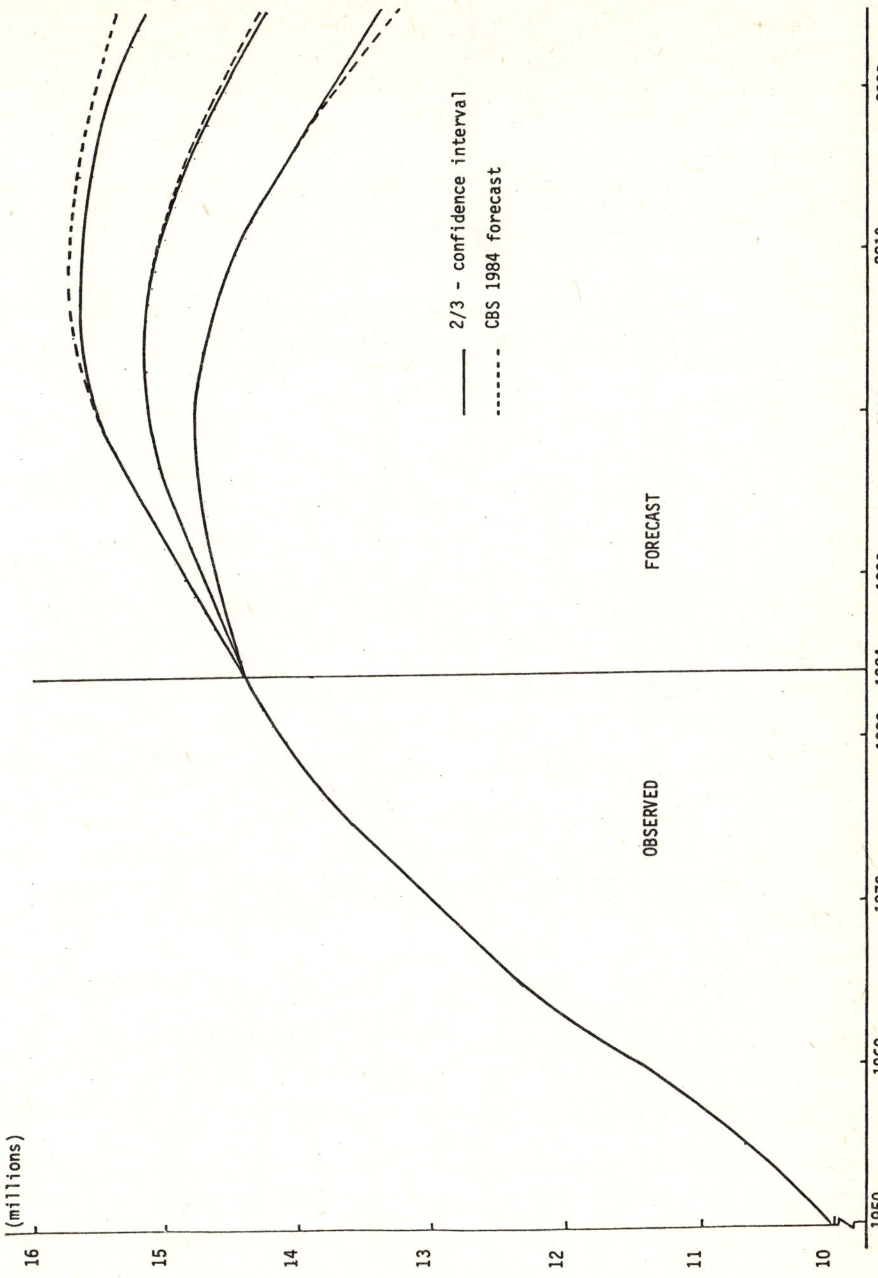


Figure 3. Population size of the Netherlands

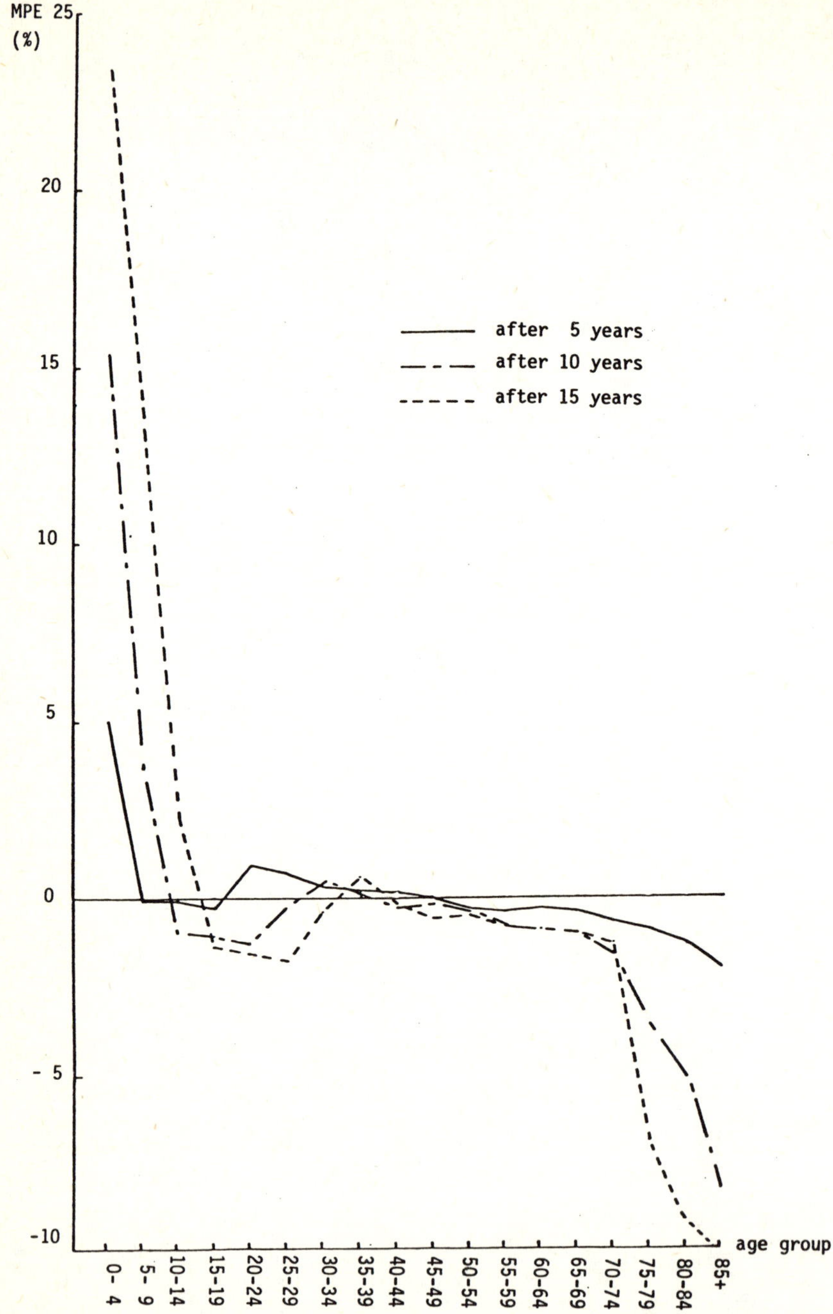


Figure 4. Mean percentage error (MPE) of post-war CBS-population forecasts, by age-group

forecast period exceeds 10 years. In Norway and Denmark similar forecast errors have been found (Leeson, p. 97; Brunborg, 1984, p. 14). It is thus relatively difficult to make a precise forecast for people over 75 years.

The relatively big forecast errors for the 0-4 year age groups call for a more accurate analysis of the reliability of fertility forecasts. To this end, a number of standards have been used: the mean error (ME), the mean absolute error (MAE), the root mean square error (RMS), the mean percentage error (MPE), the mean absolute percentage error (MAPE), and the root mean square percentage error (RMSPE). The ME, MAE and RMSE are defined in terms of the same unit of measurement as the variable to which they refer. In the ME, overestimations are compensated for by underestimations, in the MAE over- and underestimations have the same weight. The RMSE gives much weight to big errors, whereas the ME and MAE give the same weight to all errors. The MPE, MAPE and the RMSE are independent of the unit of measurement of the given variable. They can therefore be used to compare totally different situations.

In table 6 the 1950's forecasts underestimate the fertility and the post-1965 forecasts overestimate the fertility. The mean absolute error fluctuates between 2.2 thousand, that is 1.0% (1951 forecast, after 5 years) and 102 thousand, or 53.5% (1965 forecast, after 15 years). These errors are similar to those of the post-war fertility forecasts of the United States (Ahlburg, 1982). If we look at the relative (percentage) forecast errors, we see that it was just as difficult, if not more difficult, to forecast the fertility in 1970 and in 1972, than it had been in 1965. The most striking conclusion that can be drawn from table 5 is that forecast errors did not drop to a few percent until 1975. The transition from a period analysis of fertility, applied up to and including the 1965 forecast, to a cohort analysis, as from the 1970 forecast did therefore not immediately result in an increased reliability.

Table 6. Reliability of post-war CBS fertility forecasts.

| Forecastperiod | ME | MAE | RMSE | MPE | MAPE | RMSPE |
|---|---|---|---|---|---|---|
| | x 1000 | | | % | | |
| 5 years | | | | | | |
| P 1950 | -21.0 | 21.0 | 22.4 | - 9.2 | 9.2 | 9.8 |
| P 1951 | 2.2 | 2.2 | 2.6 | 1.0 | 1.0 | 1.2 |
| P 1965 | 31.2 | 31.2 | 33.0 | 12.9 | 12.9 | 13.7 |
| P 1970 | 40.6 | 40.6 | 49.4 | 20.5 | 20.5 | 25.6 |
| P 1972, A | 34.4 | 34.4 | 45.4 | 19.2 | 19.2 | 25.5 |
| P 1972, B | 23.2 | 23.2 | 28.4 | 12.8 | 12.8 | 15.8 |
| P 1975, A1 | 5.0 | 5.0 | 6.1 | 2.9 | 2.9 | 3.5 |
| P 1975, B1 | - 6.0 | 6.0 | 7.6 | - 3.4 | 3.4 | 4.3 |
| P 1980, L | 6.8 | 7.2 | 19.1 | 4.0 | 4.2 | 5.3 |
| P 1980, M | 10.6 | 10.6 | 13.3 | 6.2 | 6.2 | 7.7 |
| P 1980, H | 27.2 | 27.2 | 40.0 | 15.8 | 15.8 | 23.2 |
| 10 years | | | | | | |
| P 1950 | -27.5 | 27.5 | 29.1 | -11.8 | 11.8 | 12.4 |
| P 1951 | - 1.7 | 4.3 | 5.7 | - 0.7 | 1.8 | 2.4 |
| P 1965 | 67.4 | 67.4 | 79.9 | 31.8 | 31.8 | 39.5 |
| P 1970 | 68.4 | 68.4 | 76.5 | 37.6 | 37.6 | 42.8 |
| P 1972, A | 54.2 | 54.2 | 61.7 | 30.5 | 30.5 | 34.9 |
| P 1972, B | 32.5 | 32.5 | 35.8 | 18.3 | 18.3 | 20.2 |
| P 1975, A1 | 16.1 | 16.1 | 21.6 | 9.3 | 9.3 | 12.5 |
| P 1975, B1 | - 7.8 | 7.8 | 10.0 | - 4.4 | 4.4 | 5.6 |
| 15 years | | | | | | |
| P 1950 | -34.1 | 34.1 | 36.3 | -14.3 | 14.3 | 15.0 |
| P 1951 | - 7.6 | 9.3 | 12.2 | - 3.1 | 3.8 | 5.0 |
| P 1965 | 101.7 | 101.7 | 118.1 | 53.5 | 53.5 | 64.7 |
| P 1970 | 79.1 | 79.1 | 85.3 | 44.2 | 44.2 | 48.2 |

Rather, it was the relatively stable fertility trend after 1975 that seems to have favourably influenced the forecasting errors.

## 5. CONCLUSIONS

Demographic behaviour cannot be predicted: it is impossible to be 100% sure about the future course of fertility, mortality, migration and the like. Yet we can make a forecast: a plausible and realistic assumption of future events based on existing information. However, the reliability and durability of such forecasts are limited.

Since all forecasts have some degree of uncertainty, it is important to know more about this uncertainty. The present, deterministic approaches used for population forecasting can only express uncertainty in a qualitative manner. Several variants can be specified for the important variables, the distance between the variants giving the degree of uncertainty. Such variants may be called "uncertainty variants". They arise when qualitative social and socio-demographic hypotheses for fertility, mortality, migration and the like are translated into quantitative values for the variables belonging to these components. Variants can therefore not be interpreted as "alternative futures". This would contravene the common definition of a population forecast. The distance between the variants should not become too great: the reliability does, indeed, increase, but the amount of information given drops drastically. By increasing the margin between the variants, the growing uncertainty with time can be absorbed. In addition, the temporal uncertainty can be expressed per component with the aid of a border year, i.e., the ultimate future year for which a particular trend in fertility, mortality and migration can be assumed. Beyond this border year the given component is kept constant. The more distant the border year, the greater the certainty. In the period around the border year, a forecast gradually turns into a projection, since we do not know anything about the plausibility of hypotheses beyond these years.

By performing sensitivity analyses one can detect the most crucial input variables of the forecast. When formulating hypotheses special attention should be given to these variables - they might even be eligible for more than one variant. Once the forecast has been issued, these crucial variables should be monitored most carefully.

When presenting the results, the uncertainty should be apparent. Users of forecasts should reckon with imprecise population estimates. Decisions taken on the basis of population forecasts should either be robust against the expected forecasting errors, or else they should be flexible enough to be adjusted as soon as improved (revised) population forecasts are issued. When presenting the results, the uncertainty can be emphasized by presenting the results of the variants simultaneously. It is not advisable to incorporate the - supposedly - most plausible variant in the text itself and to insert the secondary variants in an appendix.

It is possible to gain quantitative insight into the uncertainty which is inherent to population forecasts by constructing time series models for the individual components as well as Markov-type forecasting models, and by analysing forecasting errors. Time series models are particularly useful for short-term forecasts. More attention should, in future, be given to the evaluation of existing forecasts, in an attempt to detect any regularities in the forecasting errors and to explain them. Only then can a forecast be made of the errors themselves. To this end, time series analysis may be used. This brings us a step beyond a mere explanation of the present, an explanation which will always retain a high degree of uncertainty.

## REFERENCES

Ahlburg, D.A. , How accurate are the U.S. Bureau of the Census Projections of total live births? Journal of Forecasting, 1, 365-374, 1982.

Ascher, W., Forecasting - an appraisal for policymakers and planners, The Johns Hopkins University Press, Baltimore, 1979.

Baxter, R. and I. Williams, Population forecasting and uncertainty at the national and local scale, Progress in Planning, 9(1), 1-72, 1978

Brunborg, H., Hvor sikre er befolkningsprognosene? Noen prinsipielle betraktninger om usikkerhet i befolkningsprognoser, paper presented at the Scandinavian Conference on Forecasting, Lejondal (Norway), 24-27 September, 1984.

Central Bureau of Statistics (CBS), De toekomstige demografische ontwikkeling in Nederland na 1975, Staatsuitgeverij, 's-Gravenhage, 1976.

CBS, Prognose van de bevolking van Nederland na 1980. Deel I: uitkomsten en enkele achtergronden, Staatsuitgeverij, 's-Gravenhage, 1982.

Cruijsen, H.G.J.M., Prognose van de bevolking van Nederland na 1984, Maandstatistiek van de bevolking, 3(4), 30-43, 1985.

Cruijsen, H.G.J.M. and W.D. van Hoorn, Prognose 1980 - gevoeligheidsanalyse van het rekenmodel, Maandstatistiek van de bevolking, 31(12), 20-30, 1983.

De Beer, J., A time series model for cohort data, Journal of the American Statistical Association (forthcoming) (1985).

Feichtinger, G., Stochastische Modelle demographischer Prozesse, Lexture notes in Operations Research and Mathematical Systems no. 44, Springer Verlag, Berlin, 1971.

Granger, C.W.J. and P. Newbold Forecasting Economic Time Series, Academic Press, New York,

1977.

Heisenberg, W., Nuclear Physics, Methuen and Co. Ltd., London, 1953.

Henry, L. and H. Gutierrez, Qualité des prévisions démographiques á court terme: étude de l´extrapolation de la population totale des départements et villes de France, 1821-1975. Population, 32(3), 625-647, 1977.

Hoem, J.M., Levels of error in population forecasts, Artikler fra Statistisk Sentral Byra nr. 61, Oslo, 1973.

Keilman, N.W., Bevolkingsprognose en onzekerheid, Demografie, 49, 1-4, 1983.

Keyfitz, N., Applied mathematical demography, John Wiley and Sons, New York, 1977.

Keyfitz, N. The limits of population forecasting, Population and Development Review, 7, 579-593, 1984.

Keyfitz, N., 1984, The social and political context of population forecasting, IIASA Working Paper no. WP-84-3, Laxenburg, Austria

Land, K.C., 1985, Methods for national population forecasts, a critical review, Population Research Center Paper no. 7.001, University of Texas at Austin.

Le Bras, H., 1974, Populations stables aléatoires, Population, 29, 435-464

Lee, R.D., 1974, Forecasting births in post-transition populations: stochastic renewal with serially correlated fertility, Journal of the American Statistical Association, 69, 607-617

Leeson, G.W., 1981, The elderly in Denmark in 1980: consequences of a mortality decline, European Demographic Information Bulletin, 12(3), 89-100

McDonald, J., 1979, A time series approach to forecasting Australian total live births, Demography, 16, 575-601

McDonald, J., 1981, Modeling demographic relationships an analysis of forecast functions for Australian births, Journal of the American Statistical Association, 76, 782-792

Nelissen, J. and A. Vossen, 1983, Een bevolkingsprognosemodel voor de korte termijn, Nationaal Programma Demografisch Onderzoek, Onderzoeksrapport no. 16, Voorburg.

Pflaumer, P., De Berücksichtigung der Unsicherheit bei der zukünftigen Entwicklung der Bevölkerung und der Rentenbeitrag Sätze in der Bundesrepublik Deutschland, Zeitschrift für Bevölkerungswissenschaft, 10(4), 501-530, 1984.

Pollard, J.H., On the use of the direct matrix product in analyzing certain stochastic population models, Biometrika, 53, 397-415, 1966.

Saboia, J., Modeling and forecasting population by time-series-the Swedish case, Demography, 11, 483-492, 1974.

Saboia, J., Autoregressive integrated moving average (ARIMA) models for birth forecasting, Journal of the American Statistical Association, 72, 264-270, 1977.

Scheéle, S., Osäkerhet i befolkningsprognoser, Stockholms Kommune Utrednings - och Statistikkontor, 1981.

Schweder, T., The precision of population projections studied by multiple prediction methods, _Demography_, 8, 441-450, 1971.

Stoto, M.A., The accuracy of population projections, _Journal of the American Statistical Association_, 78, 13-20, 1983.

Stoto, M.A. and A.P. Schrier, The accuracy of State Population Projections, _John Fitzgerald Kennedy School of Government Discussion Paper Series no. 117D_, Harvard University, 1981.

Sykes, Z.M., Some stochastic versions of the matrix model for population dynamics, _Journal of the American Statistical Association_, 64, 111-130, 1969.

Tennekes, H., Hoe voorspelbaar is het weer? _Intermediair_, 20(14), 17-21, 1984.

Van Dantzig, D., Voorspelling en profetie, _Statistica_ (Neerlandica), 6(4), 195-203, 1952.

# A FORECAST OF OFFICE AUTOMATION TECHNOLOGY TO 2000

Frederick A. Rossini*
Alan L. Porter

## ABSTRACT

This paper presents a forecast of developments in technologies involved in white collar automation through the year 2000. It divides the technologies according to their primary information-related function. The functions to be considered include input, processing, data base management, output, and communications. Technologies are divided into three levels of aggregation: components (e.g., microprocessors), devices (e.g., microcomputers), and systems (e.g., terminals linked to a minicomputer). The forecast is synthesized from primary and secondary sources to suggest general directions of technological developments in the office automation area.

## KEY WORDS

white collar automation, microprocessors, microcomputers, minicomputers

---

* Georgia Institute of Technology

## I. Introduction

This paper presents an overview of the projected future of technologies relating to office automation for the period to 2000. It is important to note that quantifiable parameters are most prevalent for hardware components such as microprocessors and RAM chips. For hardware devices and systems and for software, few numerical parameters exist. Our analysis will reflect these realities.

This forecast of office automation technologies will be organized by information action. These are: information production, information processing, data management, information dissemination, and interactive communication. There is no one-to one correspondence between technologies and actions. We will discuss the various technologies under the actions to which we believe they are most central. A final section will deal with systems for office automation and concluding observations.

## II. Forecasts of Technologies Supporting Information-Related Office Actions

### A. Information Production

Information production refers to the function of taking information that lies outside the office automation system and converting it to a representation within the system. At present most input is through the keyboard whose technology is essentially static.

Other input technologies assuming greater importance are optical character recognition (OCR), image processing, and voice input. Their importance lies in their directness of input, shortcircuiting intermediate steps.

OCR transforms text into electronic signals at the input of a system. The page is scanned with absorb it. The resulting image is digitalized and fed into character recognition circuitry.

In principle combining OCR with FAX (electronic transmission of images from hard copy to hard copy) is attractive since OCR involves substantial data compression. However OCR cannot yet handle graphics. When this is achieved, information compression would then allow the FAX transmission to consist of intermixed ASCII code and digitized image bits.

Further development of OCR's rests on increased speed and capability, decreased cost of hardware, and improved character recognition software. Because hardware costs are steadily falling, any bottleneck, it will be in the software. Yet, already single handwritten numbers and parts of handwritten numbers, some quite sloppily written, can be detected with an accuracy of 99% (Shridhar and Badreldin 1984). With the decrease in price and increase in capabilities of OCR equipment it is likely that we will see OCR units widely used in offices before 1990 (Kay and Powell 1983). If present trends continue, direct input of all visual material will be commonplace in the office by 2000.

The recognition of two dimensional images are within the capability of the current technology (Kanade and Reddy 1983). Such material is tackled through the brute force approach of identifying and storing the entire form with minimal compaction. As Kay and Powell (1983) note, image recognition will require the increased processing power and storage capacity that is becoming available.

In voice recognition, speech creates air pressure variations which are transduced into electrical signals which are then broken down into their component frequencies (Kaplan 1980; Doddington and Schalk 1981). In the processing system, words need to be "time aligned" to compensate for the variations in their length when they are spoken. Substantial success has been achieved in recognizing isolated words using limited vocabularies in a speaker dependent mode, in which the machine is "trained" by the speaker whose voice is to be recognized.

Continuous speech recognition is a far more complex problem since the acoustic properties of a word are modified by its place in a sentence and by adjacent words. The major current research effort in this area is by IBM. This system interprets office memos using a 5000 word vocabulary of standard English with pauses between words. An accuracy rate of 94% is reported. However, it is both speaker dependent and slow (Reddy and Zue 1983).

While there are substantial problems to overcome such as vocabulary size, speaker independence, cost and speed, and graceful error recovery, Reddy and Zue (1983) believe that systems capable of taking limited dictation might be possible in the mid 1990s if substantial resources are committed. Command recognition systems involving independent words or strings of words out of a limited vocabulary are already in use for such applications as inspections and instructions. They are now used in electronic design systems (Green 1984). Kay and Powell (1983) predict that in the late 1980s such systems may be used for voice data entry applications where a limited vocabulary is used. They also project limited continuous speech recognition systems by the mid 1990s.

B. Information Processing

In general, hardware component developments are marked by rapid increases in overall performance and in the performance/cost ratio. The key devices are computers, considered as hardware constructed in a particular architecture supported by various forms of software. Computers can be divided into three general sizes: mainframe computers, minicomputers and microcomputers. Each type plays a different significant role in office automation. Mainframes support very large data bases and large numbers of terminals. Minis can also support networks of terminals or micros simulating terminals; while micros allow desktop computing power with the possibility of being joined into networks or of acting as terminals in nets supported by mainframes or minis. Software is becoming more transparent to the user and flexible, so that the same program may be adapted to different purposes. If a bottleneck exists in processing during the period to 2000, it will be in software. Software development is presently a craft rather than engineering. Before software development will be able to keep pace with hardware development, software engineering, involving repeatability and reusability, will have to be developed to replace craft skills.

The key to hardware improvements is increased speed of processing. This ability supports the more demanding aspects of office automation, such as managing large data bases and artificial intelligence. So far the improvement has come from very large scale integration. As more capability is required for processing demands, parallel processing, the simultaneous performance of many operations in different processors is beginning to prove useful (COSEPP 1984; Davis 1983). Parallel processing was first used in the innerloops of serial machines where a small number of operations were handled simultaneously.

New parallel designs (Davis 1983) may be scalable in principle so that by adding parallel component structures without redesign, the maximum possible computing speed is increased proportionally. Some design alternatives for parallel systems (COSEPP 1984) include single-instruction multiple-data (SIMD) in which processors are given the same instruction but operate on different streams of data. In multiple-instruction multiple-data (MIMD) machines, the processors are driven by separate instruction sets, but are synchronized and able to pass data back and forth. MIMD machines are more capable of dealing with irregular patterns and complex instructions, while SIMD machines are oriented to lockstep number crunching.

Various patterns of processor connection are possible including rings, n-dimensional cubes, and binary trees. Likewise there is a tradeoff between the number and size of processors, ranging from a few very large processors to millions of single bit processors. The former approach maximizes the serial component of instructions while the latter exploits the parallelism of the instructions.

Distributed processing operating systems can take advantage of the unused processing and storage capacity of the micros on a net. In addition to distributed processing, such networking arrangements offer the potential of distributed data bases in which the data is dispersed among the various nodes in the net (Wallich 1983). While the great speed of the very large computers is not available, these systems offer large computing power shared throughout the system. They appear ideal for offices where processing capability is required locally and where information input and output takes place in many dispersed locations.

Software is an area where clearly defined parameters expressing important characteristics are lacking. There are major differences between hardware and software (Bacon 1982). First the logical complexity of software generally far exceeds that of hardware. Second, software tends to accumulate over time, while hardware is periodically replaced by improved technologies. Third, software is involved with the total workings of the institution, while hardware remains localized to the organizational component that deals directly with it.

Software is divided generically into two types: systems programs (operating systems and system wide utilities) and applications programs (Bacon 1982). Operating systems are becoming more flexible and intelligent. Multitasking in personal computer operations is advancing (Guterl 1985). Systems that deal with multi-media input and output devices will become common (Kay and Powell 1983).

In the micro-based environment, the principal current operating system families are CP/M, MS DOS, and UNIX. Kay and Powell (1983) forecast that these families will survive as competitors into the 1990s, while U.S.Department of the Treasury (1984) forecasts that UNIX-based systems will become dominant by 1990. The strength of UNIX lies in its communciations and time sharing, while it lacks transparency to the nontechnical user (Kernighan and Morgan 1982).

Applications software will emphasize the integration of different data types and functions in the same package (Chang 1983; Brown 1983). The current set of software packages, such as dBase III and Symphony, are initial efforts at the integration of programs designed for office applications. Obviously it would be desirable if one applications software package were all that was required for the individual user.

Software production will emphasize an engineering approach as opposed to the present craft approach. In a survey of software engineering practices in the U.S. and Japan, Zelkowitz et al. (1984) found little use of software engineering in either country. Present software is typically unique, without interchangeable modules. It has been described (Bacon 1982) as a bowl of spaghetti as contrasted to a set of building blocks. Thus its ability to be altered or even maintained is restricted. Software engineering techniques include rapid prototyping, the use of very high level languages, and building programs from reusable modules (Ramamoorthy et al. 1984).

Artificial intelligence (AI) is currently a popular buzzword. It refers to computerized systems that perform functions that are normally associated with "intelligence". A primary product of AI is expert systems, software packages that incorporate human expertise to deal with the complete situations that human experts confront. Current examples of expert systems include medical diagnostics, computer system configuration, and mineral exploration. In the office context AI may help in synthesizing data, composing reports, and decision support.

The key components of AI systems are knowledge bases and reasoning elements or inference engines. The knowledge bases are composed of data and insight typically associated with human experts. Knowledge acquisition is presently the principal bottleneck in constructing knowledge bases (Feigenbaum 1983). Gathering knowledge by machine using human-like strategies does not presently exist. Structuring the knowledge base is also a craft activity (Hayes-Roth 1983). Each individual case is unique; there are no general approaches. Feigenbaum (1983) predicts that heuristic approaches to knowledge acquisition should be developed by the 1990s.

Projections (U.S. Department of the Treasury 1984; Kay and Powell 1983; Feigenbaum 1983) suggest a substantial improvement in knowledge based systems by the mid 1990s. By the late 1990s these systems will become more integrated and be a central part of the operations of the more powerful computer systems and networks.

C. Data Base Management

Data base management involves mass storage devices and the software for managing the system. General trends in mass storage devices include increased storage at decreased unit cost with fewer errors. As computer networks become more sophisticated, distributed data bases should be used (Wallich 1983).

The principal mass storage media are presently magnetic disks, with tape for archival uses. Magnetic disks exist in two forms: hard and flexible. Mainframe computers presently rely on hard disk storage with disk packs in excess of a gigabyte. Microcomputers typically use 5" or 3" floppies with storage capacities in excess of 300 kbytes. Within a few years 200 mbyte 5" hard disks are likely (U.S.Department of the Treasury 1984). Access times are projected by the same source to fall from 25 msec in 1982 to 10 msec in 1997. The high capacity removable hard disk is beginning to be used more widely as a storage medium in microcomputers.

The major competitor to magnetic disks is optical disks (Goldstein 1982). They offer lower cost and higher capacity. However presently optical disks may be written on only once, and are not erasable. At costs per storage unit and access times at least an order of magnitude more favorable than magnetic disks, they are an excellent medium for archival storage (Kay and Powell 1983). With erasable optical disks anticipated by 1990, their role may be enlarged substantially. Their capacity is projected to be greater than 100 gbytes by the early 1990s (U.S. Department of the Treasury 1984; Lineback 1983).

The organization of a data base management system determines both its user friendliness and the efficiency of its performance. A tradeoff is involved between relationally structured data bases that are user friendly, slow, and require large amounts of storage, and network and hierarchical data bases that require more sophisticated users, but are more efficient and require less storage. A compromise is to interface a network system with a relational system to reap the benefits of both, but with the costs of both (Computer Decisions 1984).

The distributed data base, where the data is spread among a number of storage locations in a network, has not yet been fully perfected (Computer Decisions 1984). For example, one user can be downloading data while another is loading new data. If one node of the net goes down, the user nay be unaware that data is not being accessed. Yet the flexibility and advantages of distributed data bases are such that interest and further developments should continue.

By the late 1980s DBMS will be able to extract and summarize data for analytic and decision support. They will be able to manage geographically distributed data bases. Text, voice, and graphic information can be integrated with the data base to be accessed by query and report generation routines. By the mid 1990s users will be shielded from the data models used in the data base and will access it through high level queries and report generation routines. By the late 1990s world-wide distributed data bases will be possible (Kay and Powell 1983).

D. Dissemination

The integration of many forms of output media offers the user a wide range of options in disseminating the computer output. Image reproduction, handsome graphics, and voice mail are some of the areas where developments in output technologies will advance the current state of the art through digitizing voice and image signals and reconverting them to output form. Voice synthesizers are improving. The application of laser and fiber optics technologies has improved the process of image formation (Kay and Powell 1983).

Display technologies of interest are the CRT, the most widely used technology, and flat panels, which are becoming more widely used. The CRT is inexpensive, has moderate (512x512) resolution, and offers color capability. CRTs are being made smaller. Flat panel displays of 10 or more square inches should be available by the late 1980s. Their resolution will improve, and they may prove attractive in many applications. Interest in three dimensional imaging has generated some research. However, this technology will probably not play a major role in the timeframe of interest (Kay and Powell 1983).

Impact printers at the low end of the market are dot matrix. Improvements in cost and quality are occuring steadily. However, most new developments are in non-impact printers. Laser printing systems are capable of producing as many as 10,000 sheets per hour. Units with color capabilities should be available by the late 1980s. Ink jet technology has also been used in both high and low cost printers, and its use is becoming more widespread (Foster 1984).

Progress in speech digitalization and compression and the falling costs of voice synthesizer chips mean that this medium will be commonly available by the mid-1980s. Its main applications are in warning and status reports and in voice mail.

E. Interactive Communication

Interactive communication within the office are based on the network that transmits voice, data, text and images. The local area network (LAN) provides communications within a building or a group of buildings. The principal present alternative to the LAN is the private branch exchange (PBX). From local area to local area, long distance communications, such as phone lines or satellite relays, carry the traffic. Progress in fiber optic communications has been rapid. Standards for local and wide area networks are being developed (Bell 1985).

The primary options for local network transmission are the broadband CATV-like coaxial cable, the baseband coaxical cable, and twisted pairs of wires.

Twisted pairs of copper wires are the simplest and least costly form of connection for LANs. These are readily available and provide reliable point to point communication at speeds up to one megabit per second over a distance of one kilometer (International Data Corporation 1983).

Baseband coaxial cables offer a single high speed digital channel that can be shared by up to hundreds of users.

Bandwidths may be up to 50 mhz. depending on the type of coaxial cable used. Baseband's strengths include a simple connection that is easy to increment, a high speed channel, a well known technology, and low cost. Its weaknesses include that it can only support a single function and that voice and video are difficult because of the limited bandwidth (International Data Corporation 1983).

Broadband coaxial cables offer channels that can be used for multiple applications including voice and video. Data transmission rates in each channel are lower than baseband. However the combined data rate in all channels is higher than baseband. Broadband's strengths include well tested CATV components, 300 mhz. bandwidth available, multiple segregated channels, independent channel protocols, support of up to thousands of users. Its weaknesses include the need for a skilled support staff, potential noise interference, complex RF technology, and higher costs than equivalent baseband (International Data Corporation 1983).

There are four basic topologies of a LAN. The star has a central hub with nodes radiating from it. In the hierarchical typology, there is a subdivision of several paths for each node with nodes connecting to them and so on. A bus has a common communication path to which all nodes are connected. In the ring each node is connected to two nodes in a circle with transmission in one direction (International Data Corporation 1983).

By the end of 1982 about 16,000 LANs were installed worldwide (International Data Corporation 1983). Workstations on networks are expected to quadruple from about 200,000 in 1984 to about 800,000 in 1987 (Patterson 1984). Kay and Powell (1983) anticipate the combination of digital PBXs (private branch exchanges) with LANs by the early 1990s.

The principal alternative to the LAN is the PBX. Many potential LAN users already have PBXs. Modern, digitalized PBXs can handle both voice and data communications. Generally, LANs have greater capacity, greater expansion potential, and more speed than PBXs. However, PBX installations from a preexisting system are often cheaper (International Data Corporation 1983; see also Mayo 1982).

## III. Office Automation Systems

The purpose of this section is to integrate the technologies discussed in the previous sections into systems to illustrate the functional capabilities and potential of office automation systems to 2000. Fortunately integration does not mean connecting everything to everything. The basis of effective integration remains an understanding of the functions within the office environment. The problem of integration can be viewed from two extreme perspectives. The first is the bottom up incremental approach of interconnecting existing devices to create a system that is greater than the sum of its individual parts. This approach runs the risk of the technology driving the office functions rather than being determined by them. This incrementalism may not lead to an effective solution. The alternative approach is top down. Here the technology is imported to meet the basic functions. However this approach runs the risk that the new technology will make a major, discontinuous change in office routine that will prove disruptive (International Data Corporation 1982).

Integration has a number of dimensions. These include: physical link-up, formats and protocols, information types, and applications systems (International Data Corporation 1982).

Physical link-ups involve the physical transfer of information among devices. This is typically done electronically through a PBX or an LAN. Sometimes the transportation of information storage media serves as a link. Information exchange among different devices requires communality among coding, formats, and communications protocols. Alternatively, conversion hardware or software may be used. Different information types need to be handled in the same system. Hitherto, data and text have been the principal types. However audio, facsimile and graphic information are becoming more important. Thus future integrated systems should have the capability of handling all these forms.

Finally applications systems need to be able to integrate all these forms of information within the same applications package. This latter is a goal which has only been partially achieved in current programs.

No extant system has a full complement of the currently available functions or complete integration of the ones it has. These office systems are strongest in routine and mechanical activities and weakest in activities requiring intelligence. Decision support functions vary greatly from system to system and are not well integrated. From the managerial/professional perspective, the present systems save time that would be devoted to mechanical functions and communication, but they do not provide very much assistance in judgment and decision making. Thus they are closer to the clerical level. Yet the potential for progress based on the development of integrated packages is clearly perceived.

Between now and 2000, the individual work station will gradually evolve to the point where it will be based on powerfull 32 bit microprocessors with local removable storage of the order of high megabits or low megabits. The use of this substantial capacity will be frustrated by the relatively slower development of software capabilities. In particular the rate of the development of software possessing intelligence will increase, but not fast enough to fully exploit the hardware capabilities available before 2000. These microcomputer-based work stations will be connected in LANs, with broadband being the dominant network mode. For some applications not requiring massive, high speed data traffic, integrated voice/data PBXs will compete with the LANs. Optical fibers will become the preferred physical medium. Transmission speeds will increase substantially. Fully distributed operating systems that take advantage of the resources of the entire system to achieve more efficient processing will be more of a reality. The system iputs may be from voice, printed copy, keyboard, or possibly video. The outputs will be display, hard copy, voice, facsimile, or possibly video. A key change is that developments is artificial intelligence will render the system more transparent to the users so that most instructions can be given verbally in ordinary language or by touch.

In sum, the following areas are those where the most improvement will occur in office automation systems to 2000:

1. Massive, rapid increases in performance capability and performance/cost ratios of computer hardware;
2. Further improvement of voice, text and video related input and output components;
3. Substantial development of intelligent, transpa-

rent software;

4. Further systematic integration and enhanced communications capabilities of office automation systems.

## REFERENCES

Bacon, G., "Software", Science, 215, 775-778, 1982.

Bell, T.E., 1985, "Communications", IEEE Spectrum, January, 56-59.

Brown, M.J., 1983, "The Complete Information Management System", BYTE, December, 199-207.

Chang, D., 1983, "An Introduction to Integrated Software", BYTE, December, 103-108.

Computer Decisions, 1984, "DBMS: The More You Get The More You Want", February, 124-146 and 182-186.

COSEPP (Committee on Science, Engineering, and Public Policy, National Academy of Sciences, National Academy of Engineering, and Institute of Medicine), 1984, Report on the Research Briefing Panel on Computer Architecture, Washington, DC.

Davis, A.L., 1983, "Computer Architecture", IEEE Spectrum, November, 94-99.

Doddington, G.R., and Schalk, T.B., 1981, "Speech Recognition: Turning Theory into Practice", IEEE Spectrum, September, 26-32.

Feigenbaum, E.A., 1983, "Artificial Intelligence", IEEE Spectrum, November, 77-78.

Foster, E.S., 1984, "New Non-Impact Printers May Shake Up Stable Printer Market", Mini-Micro Systems, April 19, 113-148.

Goldstein, C.M., 1982, "Optical Disk Technology and Information", Science, 215, 862-868.

Green, D.J., 1984, "Voice Recognition: Still a Long Way to Go", Purchasing, June, 13-15.

Guterl, F., 1985, "Personal Computers", IEEE Spectrum, January, 45-49.

Hayes-Roth, F., 1983, "Codifying Human Knowledge for Machine Reading", IEEE Spectrum, November, 79-81.

International Data Corporation, 1982, Integration of Office Automation Systems, Framingham, MA.

International Data Corporation, 1983, Office Networks, Framingham, M.A.

Kanade, T., and Reddy, R., 1983, "Computer Vision: The Challenge of Imperfect Inputs", IEEE Spectrum, November, 88-91.

Kaplan, G., 1980, "Words into Action", IEEE Spectrum, June, 22-26.

Kay, P., and Powell, P., Editors, 1983, Future Information Processing Technology, 1983, Institute for Computer Sciences and Technology, National Bureau of Standards, Washington, DC.

Kernighan, B.W., and Morgan, S.P., 1982, "The UNIX Operating System: A Model for Software Design", Science, 215, 779-783.

Mayo, J.S., 1982, "Evolution of the Intelligent Telecommunications Network", Science, 215, 831-836.

Patterson, W.P., 1984, "Surviving the Micro Attack", Industry Week, June 11, 39-46.

Ramamoorthy, C.V., Prakash, A., Tsai, W.-T., and

Usuda, Y., 1984, "Software Engineering: Problems and Perspectives", Computer, October, 191-207.

Reddy, R., and Zue, V., 1983, "Recognizing Continuous Speech Remains an Illusive Goal", IEEE Spectrum, November, 84-87.

Shridhar, M., and Balredin, A., 1984, "Recognition of Isolated and Connected Handwritten Numerals", in Proceedings of 1984 IEEE International Conference on Systems, Man and Cybernetics, IEEE, New York, 142-146.

U.S.Department of the Treasury, 1984, Technology Trends and Forecasts, Washington, D.C.

Wallich, P., 1983, "Designing the Next Generation", IEEE Spectrum, November, 73-77.

Zelkowitz, M.V., Yeh, R.T., Hamlet, R.G., Gannon, J.D., and Basili, V.R., "Software Engineering Practices in the U.S. and Japan", Computer, June, 57-66.

## PART B : SOCIO-ECONOMIC IMPACT ASSESSMENT

B1 : Murdock, S.H., Leistritz, F.L. and
Hamm, R.R.
The State of Socioeconomic Impact Analysis
in the United States: an examination of
existing evidence, limitations and
opportunities for alternative futures

# THE STATE OF SOCIOECONOMIC IMPACT ANALYSIS IN THE UNITED STATES: AN EXAMINATION OF EXISTING EVIDENCE, LIMITATIONS AND OPPORTUNITIES FOR ALTERNATIVE FUTURES

Steve H. Murdock[*]
F. Larry Leistritz[**]
Rita R. Hamm[*]

## ABSTRACT

The field of socioeconomic impact analysis ha developed rapidly since the early 1970s. Despit extensive development, the field is facing a crisi with the public's environmental concern, the numbe of researchers devoting primary attention to impac analysis, and the financial support essential to th area's future productivity declining. It is thu appropriate to assess the present state of knowledg in socioeconomic impact analysis, to evaluate it conceptual and methodological limitations and t examine the likely future of the area as both a academic and a policy-relevant field of analysis This article assesses the current state of knowledg related to the economic, demographic, public servic and fiscal, and social impacts of large-scale deve lopments, discusses conceptual and methodologica limitations in the field and outlines three possibl scenarios for its future development. Its future i seen as dependent on extensive conceptual and metho dological developments and on increased sensitivit to policy-making requirements.

## KEY WORDS

socioeconomic impact analysis, present state of knowledge, impacts of large-scale developments, conceptual and methodological limitations

---

* Department of Rural Sociology, Texas Agricultural Experiment Station, Texas A&M University System

** Department of Agricultural Economics, North Dakota State University

* Department of Rural Sociology, Texas Agricultural Experiment Station, Texas A&M University System.

During the past fifteen years, an area known as socioeconomic impact analysis has become a major focus of research for numerous sociologists, economists and other social scientists.1) Although the field of socioeconomic impact analysis has progressed rapidly the area's long-term viability remains uncertain. Despite the increasing volume of recent publications in this area, public activism related to environmental issues appears to be declining (Dunlap and Dillman, 1976; Buttel and Larson, 1980), judicial support for such activities is waning (U.S. Supreme Court, 1983) and the resources necessary to produce the level of productivity experienced in the past are increasingly scarce. In like manner, recent works in the field (for example, Halstead et al., 1984; Branch et al., 1984; Freudenburg, 1984) show increasing recognition of the need to see this area of analysis as a part of the major trends in the broader disciplines. It appears possible, then, that rather than continuing to expand as a separate area of specialization, socioeconomic impact analysis may simply be subsumed by the broader disciplines with which its proponents are affiliated. Although this may appear to be a positive outcome, to many social scientists, such changes made it evident that the field is at a crossroads in its development. Consequently, this is an appropriate time to take stock of what contributions socioeconomic impact analysis has made to our knowledge of technology-induced change of what contributions it might make in the future.

At the same time, the issues of concern in socioeconomic impact analyses extend beyond disciplinary boundaries and academic concerns. Such analyses have always been part of a broad-based, multidisciplinary decision-making process (i.e., the environmental impact assessment process), and its uses for planning and management have been major foci for its development (Weber and Howell, 1982; Halstead et al., 1984). Another important reason for assessing the state of socioeconomic impact assessment is to examine how the present state of development in the field compares with the needs of the decision-making process of which it forms a part. It is essential, then, to assess the strengths and weaknesses of socioeconomic analysis as an applied area of social science, which contributes to policy formulation and implementation.

In sum, the field of socioeconomic impact analysis has developed rapidly, but is now at a critical point in its development as a specialized field of social science analysis. In addition, the fact that such analysis plays a role in policy formulation suggests that any assessment of the field also must consider its policy dimensions. It is essential, then, to evaluate the present status of socioeconomic impact analysis, both as a contributing field to the social sciences and as an aid to policy formulation and decision making.

This paper begins with a brief examination of the key questions and general findings of studies concerning several major types of socioeconomic impacts (Murdock and Leistritz, 1979). These include economic, demographic, public service and fiscal, and social impacts. The next section of the paper consists of an evaluation of conceptual and empirical limitations to the state of knowledge in socioeconomic impact analysis and of the implications of these limitations both for its scholarly development and for its utility in policy development and implementation. Finally, we suggest alternative scenarios for the future development of socioeconomic impact analysis and speculate on the viability of the field under each of these scenarios.

## SOCIOECONOMIC IMPACTS : THE PRESENT STATE OF KNOWLEDGE

In this section, we summarize the state of knowledge related to each of several major categories of impactseconomic, demographic, public service and fiscal, and social impacts. Although the categories shown are admittedly somewhat arbitrary, they have become widely recognized in the impact literature

(Murdock and Leistritz, 1979; Leistritz and Murdock, 1981), and they provide a useful basis for summarizing existing evidence concerning the socioeconomic impacts of large-scale developments. In presenting this summary, it must be noted that the evidence is drawn from analysis of large-scale resource developments in rural areas in the United States and that evidence from other types of projects in urban areas and in other nations is less adequately represented. For each of these four types of impacts, we summarize the major issues and questions that have been the focus of attention in that area of impact analysis and present available evidence from a selected number of citations concerning these impacts.

## Economic impacts

The potential economic effects of large-scale projects have been a major focus of attention because the economic impacts are perceived to be among the most positive. Although numerous aspects of economic effects have been examined (Chase et al., 1983), three have received the most attention. These include:

(1) What proportions of project construction and operational jobs are likely to be filled by local workers, as compared to immigrants and what are the likely origins of immigrating workers?

(2) What is likely to be the magnitude of the secondary (indirect and induced) employment resulting from project development? What proportions of these jobs will be filled by local workers?

(3) How will local businesses be affected by rapid growth resulting from a major project? For example, will development provide opportunities for expansion or will local firms experience difficulty competing with new chain stores and in attracting and retaining quality workers?

An examination of the characteristics of work forces associated with large-scale development projects reveals that many construction jobs are filled by relocating workers. The actual rate of nonlocal construction workers varies, however, with the characteristics of the local labor pool. Substantial variation has also been found in the hiring of operational and maintenance workers, with local hiring rates generally higher for these workers than for construction workers and higher for smaller projects and those located in more populous areas. Analysis of the origins of nonlocal workers also vary with lower proportions of local workers (from the project location state) being found for projects in sparsely populated states and for more skilled job types. The secondary employment impacts have been difficult for researchers to measure with precision. However, retrospective analysis now estimate that secondary employment effects have been smaller than previously believed (Gilmore et al., 1982; Pijawka and Chalmers, 1983; and Fookes, 1981). The areas of origin and characteristics of the workers who fill secondary jobs have received less attention than those of primary workers; however, recent studies indicate that secondary workers constitute a significant source of local growth and are not affiliated with primary workers (Halstead and Leistritz, 1983b).

Available evidence related to local economic impacts suggests that both the direct and indirect economic benefits of large-scale projects may be more modest than is often anticipated (Halstead et al., 1984) and that, like other large-scale industrial developments in rural areas (Summers et al. 1976), the economic benefits of such projects may be overestimated prior to the initiation of such projects. At the same time, however, available evidence also suggests that the expected negative economic impacts on local businesses' market shares and labor forces also may be less severe than anticipated. In sum, available evidence suggests that the local economic impacts of large-scale developments are neither inherently positive nor negative, but rather are heavily dependent on the nature of the site area and of the project.2)

## Demographic Impacts

Population growth has been one of the most publicized effects associated with large-scale developments in rural areas (Gilmore and Duff, 1975; - Leistritz and Murdock, 1981; Murdock et al., 1980). This growth has been characterized as having direct and often dramatic impacts on local communities; it is the source of many social changes and a major impetus to many types of economic changes. Thus, the extent of growth that may be expected to result from large-scale developments, the distribution of such growth among local communities and the effects that such growth is likely to have on the characteristics of the local population have been areas of paramount concern for impact researchers and policymakers (Murdock and Leistritz, 1979; Halstead et al., 1984; Murdock et al., 1982a; Murdock et al., 1983).

Existing analyses have made it apparent that the level of local demographic impacts is a function of three general sets of factors: the characteristics of the project (e.g., size, location, skill requirements, etc.); the characteristics of the local area (e.g. skill levels of local workers); and the characteristics of project-related inmigrants (e.g., their age, income, and preferences) (Murdock and Leistritz, 1979; Murdock et al., 1982a). Retrospective analyses (Gilmore et al., 1982; Chalmers et al, 1982) have established that it is the unique combination of these factors and the interaction of such characteristics that determine the magnitude and the nature of local demographic impacts. Despite considerable geographic diversity, however, generalizable information has been accumulating related to several key issues concerning demographic impacts.

Of particular concern have been the answers to

such questions as:

(1) What levels of growth have actually occurred in developing areas?

(2) What types of communities have experienced the most rapid growth?

(3) What are the characteristics of inmigrants to developing areas and how do their characteristics compare to those of longtime residents?

Overall, the available evidence in regard to the demographic impacts of large-scale developments suggests that such impacts are unlikely to result in boomtown levels of growth, but that they are difficult to predict and highly variable depending on the characteristics of the project, the local area's population and the characteristics of new inmigrants.3) The characteristics of inmigrants, however, tend to be similar to those of other migrants and thus, it appears likely that they have provided some resurgence of demographic vitality for many rural areas.

## Public Service and Fiscal Impacts

The public service and fiscal impacts of resource developments have been major areas of concern of socioeconomic impact assessments (Denver Research Institute, 1979; Murphy and Williams, 1978; Chalmers and Anderson, 1977; Leistritz and Murdock, 1981). Changes in the availability and quality of public services and the revenues and costs associated with a major project are among the most visible and widely noted impacts during any major rural development (Summers et al., 1976; Lonsdale and Seyler, 1979) and have been a topic of much discussion in the impact literature (Gilmore and Duff, 1975; Gilmore et al., 1976; Murdock and Leistritz, 1979; Halstead et al., 1984).

In these areas of impacts, the key questions have been ones such as:

(1) Which service sectors are likely to be most and which least impacted by large-scale developments?

(2) To what extent are levels of service satisfaction likely to be impacted and which services are likely to be perceived as the most negatively impacted?

(3) Do developments pay their own way in relation to public costs by generating sufficient revenue to meet the costs of increased public service demands?

In sum, the available evidence on public services suggests that service impacts have been highly variable both from site to site and by service area.4) Housing, recreation and social services, however, generally have been among the most impacted (and thus difficult to manage) and are the areas where greatest dissatisfaction has been perceived. Finally, existing evidence suggests that local areas can alter such impacts by ensuring corporate responsibility and the use of multidimensional planning strategies.

A review of recent studies concerning fiscal effects of large-scale projects suggests that the answer to the question of whether projects pay their own way must be a qualified one.5) Findings to date suggest that large, capital-intensive projects often generate revenues to state and local governments that tend to exceed total expenditures by these units over the life of the facility. However, local governments may experience substantial periods, early in the project life-cycle, in which revenues fall short of expenditures, and those local jurisdictions that are not able to directly tax the facility, but host much of the associated population, may experience chronic fiscal stress. As these characteristics of large-scale projects have become more widely understood, state and local officials and project managers have attempted to develop new institutional mechanisms, or modify existing ones, in order to alleviate the local fiscal problems associated with large projects (Halstead et al., 1984).

Thus, the existing evidence related to fiscal impacts suggests that the area is one where few overall generalizations can be drawn. Even more than in the other impact areas described above, the results in this area of analysis suggest that the impacts are areally specific. The circumstances surrounding the location and taxation status of the facility determine the magnitude and the nature of the impacts experienced. Although one might conclude, on the basis of existing evidence, that projects eventually pay for their most direct service costs, such a conclusion would likely have little generalizability to future projects occurring in different fiscal and geographic settings.

## Social impacts

The social impacts of large-scale developments often have been inappropriately defined, to include nearly all other impacts not analyzed in the economic, demographic, public service, and fiscal impact areas. Thus, social impacts often have been seen as including psychological, historical and many other dimensions that are not primarily sociological (Murdock and Leistritz, 1979). In addition, the assessment of such impacts often has been seen as a strategy to be used in gaining public acceptance of large-scale and unpopular projects (Branch et al., 1984). Even when appropriately defined to include the major social structural and social psychological dimensions basic to sociological analysis (Branch et al., 1984; Finsterbusch et al., 1983; Leistritz and Murdock, 1981), the focus of social impact analysis has been broad. Perhaps, the most commonly raised questions concerning such impacts, however, have focussed on the issues of:

(1) Do large-scale projects alter the social interaction patterns and social structural composition of rural communities?

(2) Do such projects lead to major disruptions in social control mechanisms in rural areas and thus, result in increased rates of crime, delinquency, marital dissolution, etc.?

(3) What groups are most positively and which are most negatively impacted by such projects (e.g., the elderly, the poor, the young)?

(4) What levels of social psychological stress are placed on persons living in the sitting areas of large-scale projects, and if stress is induced, does it have temporary or permanent effects on area residents?

(5) Do rural residents perceive large-scale projects as having had positive or negative impacts on their communities and which aspects do they believe have been most positively and negatively impacted?

Some of the early analyses of social impacts suggested that large-scale projects would permanently alter the interaction patterns and the social structures of rural communities, making interaction patterns less informal and leading to alterations in the sustenance and occupational bases of rural areas. Several more recent analyses have refuted such hypotheses, however. Much of the early impact literature also suggested that large-scale projects might be particularly negative for the poor, youth, women, and particularly the elderly. Recent analyses indicate that it is the youth who have experienced the greatest stress and been most negatively impacted. For nearly all other groups, however, there is little evidence of large or lasting disruptions due to rapid growth.

The question of whether large-scale and potentially dangerous projects induce stress has been a controversial area of analysis. The existing evidence is very mixed, with some researchers finding extensive levels of stress - at least for particular groups (youth) - but others finding no such effects or that such stress tends to dissipate rapidly after the peak impact period has passed. Finally, although residents' evaluations of impacts seem to be cyclical (being most positive prior to the construction of the project, decreasing during the construction stage of the project, and then increasing to high levels of support after the project has reached the operational stage), it appears that large-scale projects are seen as positive overall.6) Analyses also suggest that residents perceive economic impacts to have been the most positive and social and public service impacts to have been the most negative.

Overall, the available evidence on the social impacts of large-scale developments suggests that these impacts seldom resemble the boomtown syndrome often hypothesized in the early impact literature. Such impacts can be quite negative, but evidence suggests that they have tended to have both positive and negative impacts, with the most negative impacts being restricted to selected groups and areas. 7) Thus, analysis in this area has served to replace the boomtown stereotype with increasingly sound empirical evidence.

## CONCEPTUAL AND METHODOLOGICAL LIMITATIONS IN SOCIOECONOMIC IMPACT ANALYSIS

The previous discussion clearly shows that socioeconomic impact analysis has developed a relatively extensive body of empirical findings. Its knowledge base has increased substantially, and a considerable body of empirical research has replaced the speculation that largely characterized the field in the early 1970s. Both the conceptual and the empirical bases of the field remain limited in a number of regards, however. Before presenting our speculations concerning the future of the area, it is essential to discuss some of the major limitations that characterize the area and that restrict its development both as an area of sociale science analysis and as a tool contributing to policy formation and implementation. We begin by examining conceptual limitations and then move to a discussion of empirical and methodological limitations.

Of the major limitations in the area, perhaps none is more apparent than the lack of clear definitions for the field's basic concepts. Thus, the very concept of impacts has not been adequately delineated. Although it refers, in general, to changes resulting from a project that would not have occurred in the absence of the project (Murdock and Leistritz, 1979), the qualitative and quantitative dimensions of the concept of impacts remain nebulous. In like manner (as noted earlier), the delineation of the major types of impacts (especially, social impacts) remains ambiguous. As a result, social and other types of impacts tend to become residual categories, rather than conceptually-distinct categories. Similar problems exist in relation to such terms as secondary impacts, environmental impacts and numerous others. A case can be made that such definitional problems are no more severe than, for example, the general problems surrounding the definitions of such concepts as social change in the discipline of sociology. However, the fact that the field of socioeconomic impact analysis has yet to produce a single major work devoted to the examination of such conceptual issues indicates that the area is still in its conceptual infancy.

Closely related to the above limitation is the fact that the field still lacks a sound theoretical base and fails to adequately incorporate existing theoretical frameworks in its analyses. It is characterized by analyses that are generally theoretically uninformed. Although theoretical linkages are increasingly being made between such analyses and community theory (Branch et al., 1984), hman ecological theory (Murdock, 1979), social change and modernization theory (Freudenburg, 1984; England and Albrecht, 1984), and systems theory (Chase et al., 1983), such analyses remain the exception and are, at best, piecemeal attempts. In like manner the field is characterized by a failure to employ the best of our existing base of theoretical knowledge, even when theoretical development is attempted. Those works employing theoretical perspectives tend to use them largely as a means of justifying the reporting of empirical findings (Murdock and Schriner, 1979). Analyses attempting to develop and test theoretically grounded hypotheses are virtually nonexistent. The area has yet to produce a major work attempting to integrate the numerous dimensions of the area within a common theoretical perspective or one systematically relating aspects of its phenomenon to existing frameworks. The field is unlikely to obtain scholarly status until its theoretical bases have been developed.

In sum, then, the area of socioeconomic impact analysis must be characterized as being largely - atheoretical. Although such a characterization may be typical of a relatively new and largely inductively derived area of study, it remains the major limitation of the field at the present time.

Equally important, the lack of such theoretical development tends to limit the utility of the field for decision making and policy formulation. Because the area of socioeconomic impact analysis has failed to develop theoretically grounded generalizations, it provides decision makers facing new types of projects in previously undeveloped areas with little guidance. In order to obtain some guidance, decisionmakers must attempt to locate analyses of projects that are sufficiently similar to those they are anticipating to provide some bases for planning. A theoretically informed body of evidence is necessary to provide the bases for predicting the likely set of occurrences that may result from a large-scale project. Given the rapidly evolving technological bases of our society that promise to produce an increasing number of unprecedented developments with potentially beneficial and harmful direct and indirect effects, the lack of a sound theoretical basis for socioeconomic impact assessments not only threatens its development as an academic interest, but may make it largely irrelevant to understanding the very types of developments, the analysis of which brought it into existence.

Numerous empirical and methodological limitations are also evident (Leistritz and Murdock, 1981). Although some of these limitations, such as the lack of a sufficient number of longitudinal and multisite analyses and of evaluations of the accuracy of past predictions (Murdock et al., 1982b), are endemic to the social sciences in general, others appear to be more severe in socioeconomic impact analysis.

One of the most consequential of these limitations is the failure to develop methods to adequately integrate clearly interrelated types of impacts. Thus, numerous computerized models have been developed to quantitatively integrate economic, demographic, public service, and fiscal impacts (for an overview of such models, see Murdock and Leistritz, 1980 and Leistritz and Murdock, 1981), but social impacts are seldom integrated effectively either quantitatively of qualitatively with other dimensions. As a result, the analysis of social impacts tends to remain as an isolated enterprise in an area of analysis that requires an integrated multidisciplinary approach. Similarly, socioeconomic analyses are rarely integrated with biological or other analyses. Although such integration is admittedly difficult, given the lack of sound theoretical bases for linking various bodies of disciplinary knowledge, integrative mechanisms must be sought if the area is to be of greater pragmatic utility for policy formulation.

An additional methodological limitation is the lack of effective operational indicators for key concepts. Although this limitation is closely related to the lack of an adequate degree of conceptual development, it also presents major problems for socioeconomic impact empirical analysis and for decision makers using the results of such analysis. A prime example of this problem is that of the operationalization of the concepts of impacts and the severity of impacts. Except for the very preliminary and admittedly ad hoc delineation provided by Gilmore and Duff (1975) of 5 to 10 percent and over 10 percent population growth per year as delineating moderately and severely impacted areas, respectively, there is virtually no guidance on what various levels of impact are likely to entail for an area. In like manner, there is virtually no guidance concerning the negative or positive implications of given levels of secondary employment growth, increased service and fiscal demands or levels of community satisfaction or dissatisfaction. Until the field develops better indicators of key concepts and a more complete empirical base for establishing the likely consequences of different outcomes of such indicators, much of what is produced as a result of socioeconomic impact analyses is likely to be met with a minimum degree of enthusiasm by decision makers who wish to know the pragmatic relevance of various effects.

Finally, although much of the discussion presented in this section has stressed the need to more effectively integrate the field of socioeconomic impact analysis with the theoretical bases of social science disciplines, it is equally important to stress that the field's birthplace of pragmatic concerns should not be ignored. The research questions investigated, and the methods used, must produce results that have policy relevance as well as academic significance. Too often, the field has been characterized by analyses that pursue one of these goals, at the expense of the other. For example, social scientists may collect data to address academic concerns and consultants may collect data to address the immediate informational needs of decision makers. As a result, the acquisition of data that can be used both to extend the base of substantive knowledge and address practical concerns is often foregone. Multipurpose research designs should be employed to ensure that both pragmatically and conceptually useful results are obtained.

The field of socioeconomic impact analysis requires additional conceptual and methodological development. Its present empirical and methodological weaknesses are apparent, and promise to limit its policy relevance. Thus, refinement in these areas is essential to the future development of the field.

## THE FUTURE OF SOCIOECONOMIC IMPACT ANALYSIS

In the two preceding sections, we have attempted to delineate the present state of knowledge in the field and to discuss some of the major conceptual, empirical and methodological weaknesses that exist in the field at its present stage of development. In this final section, we wish to suggest possible alternative futures for the area of socioeconomic impact analysis. Although any such effort is likely to have limited success, and perhaps to be of limited utility, it is in the nature of socioeconomic impact analysis to provide such scenarios as a guide for choosing (or at least moderating) the future one wishes to achieve.

One potential future could involve the dissolution of the area of socioeconomic impact analysis, at least as a distinct area of academic interest. As noted in the introduction, interest in the area, the types of large-scale developments that initially fostered interest in it and the financial resources to support it appear to be diminishing. It is possible, then, that academics interested in the area will simply pursue other interests or come to couch their analysis as specialized analysis in existing branches of the social sciences. Because occurences of large-scale, technologically based developments are likely to continue, such a scenario may result in the area remaining in existence, but largely as the province of consultants completing definitive pragmatic administrative requirements. Such a dualistic evolution of the field has parallels in engineering, but would be unfortunate. Academics would los a valuable laboratory for assessments of the implications of key socioeconomic processes and decision makers would lose the potential for obtaining useful, theoretically informed, predictive tools for anticipating such impacts in the future.

A second possible scenario, and perhaps the most likely, is a continuation of the status quo. This, too, has implications for the field. Under this scenario, the number of analysts with major interests in the area would continue to decline and although their contributions would be crescive, the field would progress slowly. Additionally, under this scenario, the theoretical, empirical and methodological bases of the area are also likely to develop very slowly and the status of the area as one with tenuous academic credibility is likely to continue.

Equally important, the limited utility of the field for policy analysis would continue. In fact, continuation of the present limited utility of the area for policy analysis could result in a situation that is the reverse of that forecast under the first scenario. That is, social scientists may continue to pursue the field as an area of academic interest, but decision makers may come to conclude that the area of socioeconomic impact analysis is of little relevance to their decisionmaking process. A situation in which the field is of "only" academic interest could evolve. Thus, we believe that the present status of the field may be untenable, if it is to continue to be both an academic and a policy relevant area of analysis.

A third, and perhaps overly optimistic, scenario can be forecast. This is one in which the area builds on its base of existing knowledge, addresses its limitations effectively enough to evolve into a permanent, contributing field in the social sciences and is attentive enough to the needs of decision makers to retain recognition as an area with policy relevance and utility. Despite its apparent positive features this scenario should not be seen as an "ideal" one. It is unclear, for example, whether the disciplines involved are better or less well served by the proliferation of separate specialties, and the contributions of social scientists to policy formulation have yet to be established as benevolent ones. Neither is it unproblematic to discern what events might provide an impetus for the actualization of this scenario, in a period of limited resources. Many of the aspects of this scenario, however, do not require additional professional or financial resources. They require, instead, a greater degree of self-examination by professionals in the area and an establishment of new priorities aimed at developing the empirical and conceptual basis of the field.

Clearly, these scenarios are not the only ones that can be envisioned nor are their descriptions exhaustive of the forms of change that may occur in the area of socioeconomic impact analysis in the years to come. They admittedly reflect our biases, concerns and preferences. We are not confident of which of these may occur in the future because elements of each can be seen in existing trends. Thus, the present decline in the financial support for such analysis seems likely to reduce the number of academics studying such impacts and to expedite the events described in the first scenario. At the same time, the decrease in governmental concern in the U.S. for the environment seems likely to result in the treatment of compliance with NEPA as largely an administrative requirement (reminiscent of the second scenario), and the recent publication of more theoretically informed works in the area suggests that elements of the third scenario may be emerging.

It is not clear to us, then, what, if any, longterm future may exist for the field of socioeconomic impact analysis. We obviously hope, that the area will continue to have viability and expand to have the dimensions of the most positive scenario. If it does not, we believe social scientists in the United States will have lost a unique opportunity to be involved in a form of analysis that can contribute both to the development of a more complete base of knowledge for the social sciences and to make an important contribution to a more informed process of public policy formulation.

## NOTES

1) The disciplinary and theoretical bases for the development of the field were, of course, in existence long before the 1970s. If seen as a specialized form of social change analysis (Burdge and Opryszek, 1981), community analysis (Branch et al., 1984), as an analysis of rural development or industrialization (Summers et al., 1976; Summers and Selvik, 1982; Lonsdale and Seyler, 1979), or as a form of environmental or ecological analysis (Murdock, 1979, the roots of the area are as old as the disciplines in which they are based.

   It appears, in fact, that the degree of specialization in the area of socioeconomic impact analysis since 1970 has resulted from several concomitant occurrences. First, special sociopolitical circumstances (e.g., the energy crisis) intensified the need for large-scale developments in rural areas and increased the resources available for analysis of their impacts (Murdock and Leistritz, 1979). Second, a societal-wide recognition that technology application could be detrimental to man's health and welfare emerged (Rodgers, 1976). Finally, during the same period of time (i.e., the 1970s), there was a substantial increase in the number of sociologists and other social scientists seeking to establish a specialized professional niche for themselves. In examining the state of socioeconomic impact analysis, then, it is essential to remember both that its origins resulted from a unique set of sociohistorical events and that the issues it addresses are basic to existing disciplines.

2) A review of numerous studies was the basis of this evaluation; see especially, Mountain West Research, 1975; Chalmers, 1977; Dunning, 1981; -DeVeney. 1977; Malhotra and Manninen, 1980; Pearson, 1984, Leistritz et al., 1982; Wieland et al., 1977; Hooper and Branch, 1983; Browne, Bortz and

Coddington, Inc., 1981 and 1982; Baker, 1977; Murdock and Leistritz, 1979; Gilmore et al., 1982; Pijawka and Chalmers, 1983; Fookes, 1981; Berkey et al., 1977; Denver Research Institute, 1979; Halstead and Leistritz, 1983a and 1983b; Halstead et al., 1984; and Summers et al., 1976.

3) Among the studies included in this review of demographic analyses were: Murdock and Leistritz, 1979; Murdock et al., 1982a and 1982b; Gilmore et al., 1982; Chalmers et al., 1982; Summers and Selvik, 1982; Murdock et al., 1980; Murdock et al., 1984a and 1984 b; Mountain West Research,1975; Wieland et al, 1977; Halstead et al., 1984; and Ritchey, 1976.

4) Public service impacts cited in the following studies, among others, provide the basis for this conclusion: Leistritz and Murdock, 1981; Metz, 1980 and 1982a; Gilmore et al., 1979 and 1982; Halstead and Leistritz 1983a; Leistritz and Maki, 1981, Dixon, 1978; Davenport and Davenport, 1979; Wilkinson et al., 1982; Uhlman and Olson, 1984; Schriner et al., 1976; Murdock and Schriner, 1979; England and Albrecht, 1984; Albrecht et al., 1985; Weber and Howell, 1982; Halstead et al., 1984; ROMCOE, 1982; and Pinfield and Etherington, 1982.

5) Consideration was given to the fiscal impacts cited in numerous studies, including Gilmore et al., 1976; Murdock and Leistritz, 1979; Halstead et al., 1984; Murray and Weber, 1982; and Leistritz and Murdock, 1981.

6) A recent analysis by Freudenburg and Baxter (1984), however, suggests that perceptions of large-scale nuclear projects may have been permanently altered by the accident at Three-Mile Island such that residents remain skeptical about the benefits of such projects.

7) Among the studies dealing with social impacts of

large-scale developments reviewed for this analysis are: Gilmore and Duff, 1975; Murdock and Schriner, 1978; England and Albrecht, 1984; Murdock et al., 1981 and 1983; Gilmore et al., 1982; Chalmers et al., 1982; Davenport and Davenport, 1979; Wilkinson et al., 1982; Freudenburg et al., 1982; Albrecht, 1982; Lantz and McKeown, 1977; Murdock and Leistritz, 1979 and 1982; Freudenburg, 1982 and 1984; Albrecht et al., 1985; Freudenburg and Jones, 1984; Krannich et al., 1984; Sills et al., 1982; Fritz and Marks, 1954; Dynes, 1974; Weisz, 1979; Rossi et al., 1978; Thompson and Blevins, 1983; and Gilmore and Duff, 1975.

## REFERENCES

Albrecht, D.E., 1985, S.H. Murdock, J. Halstead, F.L. Leistritz, and S. Albrecht "The impacts of large-scale developments on rural communities in the Western United States". In H.K. Schwarzweller and F.A. Fear (eds.), Research in Rural Sociology and Development. Greenwich, CT: Jai Press

Albrecht, S.L. 1982, "Commentary´. Pacific Sociological Review 25 (3): 297-306

Baker, J.S., 1977, Labor Allocation in Western Energy Development. Salt Lake City: University of Utah, Human Resources Institute Monograph #5

Berkey. E., N.G. Carpenter, W.C. Metz, D.W. Meyers, D.R. Portes, J.E. Singley, and R.K. Travis, 1977, Social Impact Assessment, Monitoring, and Management by the Electric Energy Industry Pittsburgh, PA: Energy Impact Associates, Inc.

Branch, K., D.A. Hooper, J. Thompson, and J. Creighton, 1984, Guide to Social Assessment: A Framework for Assessing Social Change. Boulder, CO: Westview Press

Browne, Bortz, and Coddington, Inc., 1981, A Retrospective Analysis of the Jim Bridger Complex Socioeconomic Monitoring Report. Denver: Browne, Bortz, and Coddington, Inc.,

Browne, Bortz, and Coddington, Inc., 1982, The Campbell County Monitoring Program. Denver: Browne, Bortz, and Coddington, Inc.

Burdge, R.J., and P. Opryszek, 1981, Coping with Change: An Interdisciplinary Assessment of the Lake Shelbyville Reservoir. Urbana, IL: Institute for Environmental Studies, University of Illinois.

Buttel, F.H., and O.W. Larson, III, 1980, "Whither environmentalism? The future of political paths of the environmental movement". Natural Resources Journal 20(2): 323-344

Carley, M.J., and E.S. Bustelo, 1984, Social Impact Assessment and Monitoring: A Guide to the Literature, Boulder, CO:Westview Press.

Chalmers, J.A., 1977, Construction Worker Survey, Denver: U.S. Bureau of Reclamation.

Chalmers, J.A., and E.J. Anderson, 1977, Economic-Demographic Assessment Manual: Current Practices, Procedural Recommendations, and a Test Case. Denver: U.S. Bureau of Reclamation

Chalmers, J.A., D. Pijawka, K. Branch, P. Bergman, J. Flynn, and C. Flynn, 1982, Socioeconomic Impacts of Nuclear Generating Stations, Washington, D.C.: Government Printing Office, U.S. Nuclear Regulatory Commission, NUREG/CR-2750

Chase, R.A., F.L. Leistritz, and J.M. Halstead, 1983, "Assessing the economic and fiscal effects of repository development.", in S.H. Murdock, F.L. Leistritz, and R.R. Hamm (eds.), Nuclear Waste: Socioeconomic Dimensions of Long-Term Storage, Boulder,

CO: Westview Press.

Davenport, J.A., and J. Davenport, III (eds.), 1979, Boom Towns and Human Services Laramie, WY: University of Wyoming, Department of Social Work.

Denver Research Institute, 1979, Socioeconomic Impacts of Western Energy Resource Development, Washington, D.C.: Council of Environmental Quality.

DeVeney, G.R., 1977, Construction Employee Monitoring, Knoxville, TN: Tennessee Valley Authority

Dixon, M. 1978, What happened to Fairbanks? The Effects of the Trans-Alaska Oil Pipeline on the Community of Fairbanks, Alaska, Boulder, CO: Westview Press

Dunlap, Riley E., and Don A. Dillman, 1976, "Decline in public support for environmental protection: evidence from a 1970-1974 panel study." Rural Sociology 41: 382-390

Dunning, C.M., 1981, Report of Survey of Corps of Engineers Construction, Fort Belvoir, WA: U.S. Army Corps of Engineers, Institute for Water Resources Research Report 81-R05

Dynes, R.R., 1974, Organized Behavior in Disaster, Lexington, MA: D.C. Heath and Co.

England, J.L., and S.L. Albrecht, 1984, "Boomtowns and social disruption", Rural Sociology 49(2): 230-246

Finsterbusch, K., and C.P. Wolf (eds.), 1977, Methodology of Social Impact Assessment. 1st ed. Stroudsburg, PA: Dowden, Hutchinson, and Ross, Inc.

Finsterbusch, K., and C.P. Wolf (eds.), 1981, Methodology of Social Impact Assessment, 2nd ed. Stroudsburg, PA: Dowden, Hutchinson, and Ross, Inc.

Finsterbusch, K., L.G. Llewellyn, C.P.Wolf (eds.), 1983, Social Impact Assessment Methods, Beverly Hills, CA: Sage Publications.

Fitszimmons, S.J., L.I. Stuart, and P.C. Wolff, 1977, Social Assessment Manual, Boulder, CO: Westview Press

Fookes, T.W., 1981, Expectations and Related Findings, 1973-81, Hamilton, New Zealand: University of Waikato

Fritz, C.E., and E.S. Marks, 1954 "The NORC studies of human behavior in disaster', The Journal of Social Issues 10(3): 26-41

Freudenburg, W.R., 1982, "Balance and bias in boomtown research", Pacific Sociological Review 25

Freudenburg, W.R., 1984, "Differential Impacts of rapid community Growth", American Sociological Review 49(5): 697-705

Freudenburg, W.R., and T.R. Jones, 1984, "Does an unpopular facility cause stress? A test of the Supreme Court hypothesis", Scientific Paper no. 6873, Research Project 0478, Agricultural Research Center, Washington State University.

Freudenburg, W.R. and R.K. Baxter, 1984, "Host community attitudes toward nuclear power plants: a reassessment", Social Science Quarterly 65 (4): 1129-1136

Freudenburg, W.R. and E.A. Rosa (eds.) 1984 Public reactions to nuclear Power: are there critical masses?, Boulder, CO, Westview Press

Freudenburg, W.R., L. Bacigalupi, and C. Young, 1982 "Mental health consequences of rapid growth: a report from the longitudinal study of boom town mental health impacts". Journal of Health and Human Resources Administration Winter: 334-352

Gilmore, J.S., and M.K. Duff, 1975, Boom town Growth Management: A Case Study of Rock Springs-Green River, Wyoming, Boulder, CO: Westview Press

Gilmore, J.S., K.D. Moore, and D.M. Hammond, 1976, Synthesis and Evaluation of Initial Methodologies for Assessing Socioeconomic and Secondary Environmental Impacts of Western Energy Resource Development, Denver: Denver Research Institute, WP no. 2

Gilmore, J.S., D.M. Hammond, K.D. Moore, J. Johnson, and D.C. Coddington, 1982, Socioeconomic Impacts of Power Plants, Denver: Denver Research Institute, Report prepared for Electric Power Research Institute.

Halstead, J.M., and F.L. Leistritz, 1983a, Impacts of Energy Development on Mercer County, North Dakota, Fargo, ND: North Dakota Agricultural Experiment Station, Agricultural Economics Report No. 170

Halstead, J.M., and F.L. Leistritz, 1983b, Impacts of Energy Development on Secondary Labor Markets, Fargo, ND: North Dakota Agricultural Experiment Station, Agricultural Economics Report no. 178

Halstead, J.M., R.A. Chase, S.H. Murdock, and F.L. Leistritz, 1984, Socioeconomic Impact Management: Design and Implementation, Boulder, CO: Westview Press.

Hooper, J.E., and K.M. Branch, 1983, Big Horn and Decker Mine Worker Survey Report, Billings, MT: Mountain West Research-North, Inc.

Kohrs, E., 1974, Social Consequences of Boom Growth in Wyoming, Casper, WY: Central Wyoming Counseling Center

Krannich, R.S., B. Golesorkhi, and T. Greider, 1984, "Rapid growth and personal stress: an assessment of stress levels, network ties and social support in energy-impacted communities". Paper presented at the annual meetings of the Rural Sociological Society, College Station, TX

Lantz, A., and R. McKeown, 1977, Rapid Growth and the Impact of Quality of Life in Rural Communities: A Case Study. Denver: Denver Research Institute, University of Denver.

Leistritz, F.L., and K.C. Maki, 1981, Socioeconomic Effects of Large-Scale Resource Development Projects in Rural Areas: The Case of McLean County, North Dakota. Fargo, ND: North Dakota Agricultural Experiment Station, Agricultural Economics Report No. 15

Leistritz, F.L.,and S.H. Murdock, 1981, Socioeconomic Impact of Resource Development: Methods for Assessment, Boulder, CO: Westview Press

Leistritz, F.L., S.H. Murdock, and A.G. Leholm, 1982, "Local economic changes associated with rapid growth", Pp. 25-62 in B.A. Weber and R.E. Howell (eds.): Coping with Rapid Change in Rural Communities, Boulder, CO: Westview Press

Lonsdale, R.E., and H.L. Seyler (eds.), 1979, Nonmetropolitan Industrialization, New York: John Wiley and Sons

Malhotra, S., and D. Manninen, 1980, Migration and Residential Location of Workers at Nuclear Power Plant Construction Sites. Vols. 1 and 2. Seattle, WA: Battelle Human Affairs Research Center.

McKell, C.M., D.G. Browne, E.C. Cruze, W.R. Freudenburg, R.L. Perrine, and F. Roach (eds.), 1984, Paradoxes of Western Energy Development: How Can We Maintain the Land and the People if We develop?Boulder, CO: Westview Press, AAAS Selected Symposium 94

Metz, W.C., 1980, "The mitigation of socioeconomic impacts by electric uitlities" Public Utilities Fortnightly 106

Metz, W.C., 1982a, "American energy and mineral industry involvement in housing", Minerals and the Environment, 4: 131-140

Metz, W.C., 1982b, "Energy Industry involvement in worker transportation", Transportation Quarterly 36 (4): 563584

Metz, W.C., 1983, "Industry initiatives in impact mitigation" In Proceedings of the Alaska Symposium on the Social, Economic, and Cultural Impacts of Natural Resource Development, Fairbanks: University of Alaska

Mountain West Research, Inc., 1975, Construction Worker Profile, Washington D.C.: Old West Regional Commission

Murdock, S.H., 1979, "The potential role of the ecological framework in impact analysis", Rural Sociology 44

Murdock, S.H., and F.L. Leistritz, 1979, Energy Development in the Western United States: Impact on Rural Areas, New York: Praeger Publishers

Murdock, S.H., and F.L. Leitritz, 1980, "Selecting socioeconomic assessment models: a discussion of criteria and selected models". Journal of Environmental Management 10(1): 1-12

Murdock, S.H., and F.L. Leistritz, 1982, "Commentary" Pacific Sociological Review 25 (3): 357-366

Murdock, S.H., and E.C. Schriner, 1978, "Structural and distributional factors in community development", Rural Sociology 43(3): 426-449

Murdock, S.H., and E.C. Schriner, 1979, "Community service satisfaction and stages of community development: an examination of evidence from impacted communities", Journal of the Community Development Society 10(1): 109-124

Murdock, S.H., F.L. Leistritz and E.C. Schriner, 1980. "Migration and energy developments: implications for rural areas in the Great Plains". In D. Brown and J. Wardwell (eds.)New Directions in UrbanRural Migration, New York: Academic Press

Murdock, S.H., F.L. Leistritz, and E.C. Schriner, 1982a "The demographic impacts of rapid economic development", Pp 63-96 in B.A. Weber and R.E. Howell (eds.) Coping with Rapid Change in Rural Communities, Boulder, CO: Westview Press

Murdock, S.H., F.L. Leistritz, and R.R. Hamm, 1983, Nuclear Waste: Socioeconomic Dimensions of Long-Term Storage, Boulder, CO: Westview Press

Murdock, S.H., F.L. Leistritz, R.R. Hamm, and S.S. Hwang, 1982b, "An assessment of socioeconomic assessments: utility, accuracy and policy considerations." Environmental Impact Assessment Review 3(4): 333350

Murdock, S.H., F.L. Leistritz, R.R. Hamm and S.S. Hwang, 1984a "An assessment of the accuracy and utility of socioeconomic impact assessments." in C.M. McKell, D.G. Browne, E.C. Cruze, W.R. Freudenburg, R.L. Perrine, and F. Roach (eds.) Paradoxes of Western Energy Development. Boulder, CO: Westview Press

Murdock, S.H., J. de Montel, F.L. Leistritz, P. Hopkins, and R. Hamm, 1981, An analysis of the Construction Impacts of Coal Development in Rural Texas: The case of Fayette County, Texas. Report no. 81-2, College Station, TX: Texas Agricultural Experiment Station

Murdock, S.H., F.J. Leistritz, R.R. Hamm, S.S. Hwang, and B. Parpia, 1984b,"An assessment of the accuracy of a regional economic-demographic projection model", Demography 21(3):383-404

Murphy and Williams, Consultants, 1978, Socioeconomic Impact Assessment: A Methodology Applied to Synthetic Fuels, Washington, D.C.: U.S. Department of Energy.

Murray, J.A., and B.A. Weber, 1982, "The impacts of rapid growth on the provision and financing of local public services". Pp. 97-114 in B.A. Weber and R.E. Howell (eds.), Coping with Rapid Growth in Rural Communities, Boulder, CO: Westview Press

Pearson, C., 1984, ITAT Construction Work Force Report, Bismarck, ND: Inter-Industry Technical Assistance Team (Basin Electric Power Corp.)

Pijawka, D., and J.A. Chalmers, 1983, "Impact of nuclear generating plants on local areas" Economic Geography 59(1):66-80

Pinfield, L.T., and L.D. Etherington, 1982, Housing Strategies of Resource Firms in Western Canada, Burnaby, British Columbia: Simon Fraser University, Department of Business Administration

Ritchey, P.N., 1976, "Explanations of migration". The Annual Review of Sociiology II, Palo Alto, CA: Annual Review, Inc.

Rodgers, J.L., 1976, Environmental Impact Assessment, Growth Management and the Comprehensive Plan, Cambridge, Massachusetts: Ballinger Press

ROMCOE, 1982, Rapid Growth Communities Project: Financing, Experiences, and Advise, Boulder, CO:

Rocky Mountain Center on Environment, Center for Environmental Problem Solving

Rossi, P.H., J.D. Wright, S.R. Wright, and E. Weber-Burdin, 1978, "Are there long term effects of American natural disasters?" Mass Emergencies" 3: 117-132

Schriner, E.C., J. Query, T.McDonald, and F. Keogh, 1976, An Assessment of the Social Impacts Associated with a Coal Gasification Complex Proposed for Dunn County, North Dakota, Fargo, ND, North Dakota State University

Sills, D.L., C.P. Wolf, and V.B. Shelanski (eds.), 1982, Accident at Three Mile Island: The Human Dimensions, Boulder, CO: Westview Press

Summers, G.F., and A. Selvik (eds.), 1982, Energy Resource Communities, Madison, WI: MJM Publishing Co.

Summers, G.F., S.D. Evans, F. Clemente, E.M. Beck, and J. Minkoff, 1976, Industrial Invasion of Nonmetropolitan America: a Quarter Century of Experience, New York: Praeger Publishers

Thompson, J.G., and A.L. Blevins, 1983, "Attitudes toward energy development in the Northern Great Plains", Rural Sociology 48 (1)

Uhlman, J.M., and J.K. Olson, 1984, Planning for Rural Human Services, Denver: Office of Human Development Services, Department of Health and Human Services

U.S. Supreme Court, 1983, Metropolitan Edison Co. vs. People Against Nuclear Power (PANE), 103 S.Ct. 1556, Washington, D.C.

Weber, B.A., and R.E. Howell (eds.), 1982, Coping with Rapid Growth in Rural Communities, Boulder,

CO: Westview Press

Weisz, R., 1979, 'Stress and mental health in a boom town" in J.A. Davenport and J. Davenport (eds.) Boomtowns and Human Services, Laramie, WY: University of Wyoming, School of Social Work

Wieland, J.S., F.L. Leistritz, and S.H. Murdock, 1977, Characteristics and Settlement Patterns of EnergyRelated Operational Workers in the Northern Great Plains, Fargo,ND: North Dakota Agricultural Experiment Station, Agricultural Economics Report No.123

Wieland, J.S., F.L. Leistritz, and S.H. Murdock, 1979, "Characteristics and residential patterns of energyrelated work forces in the Northern Great Plains", Western Journal of Agricultural Economics 4

Wilkinson, K.P., J.G. Thompson, R. Reynolds, Jr., and L.M. Ostresh, 1982, "Local social disruption and western energy development: a critical review". Pacific Sociological Review 25(3): 275296

Wilkinson, K.P., R. Reynolds Jr., J.G.Thompson and L.M. Ostresh, 1983, "Divorce and recent net migration into the Old West", Journal of Marriage and the Family, 45: 437445

Wilkinson, K.P., R. Reynolds Jr., J.G. Thompson and L.M. Ostresh, 1984, "Violent crime in the western energy development region", Sociological Perspectives 27(2):24125

## PART C : SOCIAL IMPACT ASSESSMENT

# SOCIAL IMPACT ASSESSMENT BY SOCIAL SIMULATION

P. Boskma*

## ABSTRACT

Social simulation is a methodology, in which real social interaction is generated under simulated conditions, representing the reference system. Models have been developed for production organizations, science policy advisory councils and governmental bureaucracies. In more than 50 experiments with over 1000 participants the psychological, structural and process validity of the models have been studied. A plurality of distinct social and cognitive frameworks can be generated and reproduced systematically. Moreover it is possible to realize policy interventions (c.q. strategies for change) within the social reality of the experiments. Thus the models can be used both for social impact assessment and as a learning strategy for the development of policy options and strategies for social change.

## KEY WORDS

social simulation, social interaction, policy interventions, strategies for change.

* Twente University of Technology
Enschede, Netherlands

## 1. INTRODUCTION

In this article I will discuss in how far the method of social simulation can be useful for social impact assessment. I will define social impact assessment as the study of the social effects of technological development and policy interventions in complex social configurations. Such policy interventions include diverse activities, such as governmental policies, activities of action groups, strategies of trade-unions and management strategies in organizations. The social effects can be quite diverse too: a change of social structures, new patterns of cooperation and conflict, existing problems loosing their saliency and new issues appearing on the agenda, shifts in cultural attitudes and value orientations, etc.

An important aspect of social impact assessment, if it is to be policy relevant, is that it can provide insight in the social effects of policy alternatives on one hand and that it can help in discovering policy options and strategies on the other hand. Its methodologies need to be appropriate to these objectives.

My main thesis is that the method of social simulation in which a group of participants generate a social situation under specified conditions, represents such a methodology, because:

1. it provides options for experimenting in real social situations (psychological validity) with major social and cognitive characteristics of real life situations (structural validity; process validity);
2. the experiments are reproducible and thus effects of different conditions can be systematically studied;
3. it is possible to generate by social simulation a plurality of social realities with their own coherencies and social structures;
4. it is possible to realize policy interventions within the social reality of the experiments

and to study their effects;
5. it is possible to investigate systematically the influence of differences in cultural attitudes;
6. social simulation experiments represent a learning strategy for the development of policy options.

I do not pretend social simulation to be the methodology for social impact assessment, of course. Its major advantage is that social impact assessment can be studied under reproducible conditions in a real social environment. To my knowledge, there is no alternative for that, so far. In my concluding remarks I will go deeper into its limitations and attractions. Before doing so, I will try to explain what the method of social simulation is and elaborate on the 6 statements above on the basis of our own research.

## 2. SOCIAL SIMULATION

Quite a diverse spectrum of social simulation models have been developed during recent decades 1). This is not the place to review them. The methodology is still quite pluralistic and in a phase of development. That is the main reason that I will concentrate on the type of models and experiments we have designed and used in our own research. Van der Meer 2) reports on 24 systematic experiments with a social simulation of a production organization in which strategies for social change are enacted. Wilhelm and Bolk 3) report on a model of the Dutch science policy advisory councils ("sectorraden" in Dutch) in which the effectivity of such councils as a function of their advisory strategies have been systematically and heuristically studied in 14 experiments. This model has been developed at the request of our Ministry of Education. A third model of a governmental bureaucracy has been developed at the request of our Ministry of the Interior, focusing on problems of delegation of tasks and local autonomy within governmental departments. So far three experiments, two of them with civil servants as participants, have taken place.

The experiments take a period of three days, two days for the experiment itself and one day for an evaluative discussion. The number of participants in the model of the production facility is 25-30, in the science policy advisory council 35-40 and in the governmental bureaucracy around 40.

It is very important to understand that the method of social simulation does not mean a simulation of social processes; it means real social interaction under simulated conditions. These conditions represent a reference system, usually a social situation as e.g. the production facility. Participants get a description of the situation and an instrumentation in the form of material objects, spatial arrangements, information, etc. This is usually called the simulation model. In the model of the production facility we designed the situation of a construction enterprise, where real, but small, houses have to be constructed. Raw materials and production apparatus have to be bought and houses can be sold on the market. Basic to the models is that such elementary processes generate a material basis of the situation. The model describes this in terms of interlocking tasks and positions. Fig. 1 shows the network of tasks and positions.

Participants are invited to run the facility and to handle it according to their own ideas of preferences with regard to labor relations. They can either continue or change the situation. The rules of the experiment are very few and simple: (1) Time goes by. It is structured to a flow sheet in simulation years of 80 minutes in newtoninan time, but time cannot be stopped, (2) a major part of the communication between subgroups is in writing, although everybody can make an appointment with any other participant for a face-to-face meeting, if both parties agree on that, (3) the participants have to earn their own living (i.e. meals and drinks) during the two days of the experiment (about 10 years in simulation time).

Although positions are indicated with convential names like personnel officer, worker and manager, participants are not asked to play a role, as is the case in many role-playing experiments. Roles emerge in interaction between the participants, using the cues presented in the model and in a process of interlocking behaviour. Thus the experiment generates its own social reality.

We developed systematic procedures for observation and questionnaires at different moments during the experiments. Thus one can study both the development process and the characteristics of actual situations.

## 3. SOCIAL SIMULATION AS A METHODOLOGY FOR SOCIAL IMPACT ASSESSMENT

Let we now turn to my six theses and report about the results we got in our own research. I am not claiming that for all types of social simulation models these conclusions hold. For many models the problem has not seriously been studied. I do claim that it is possible to design quite a range of models for which our results are valid.

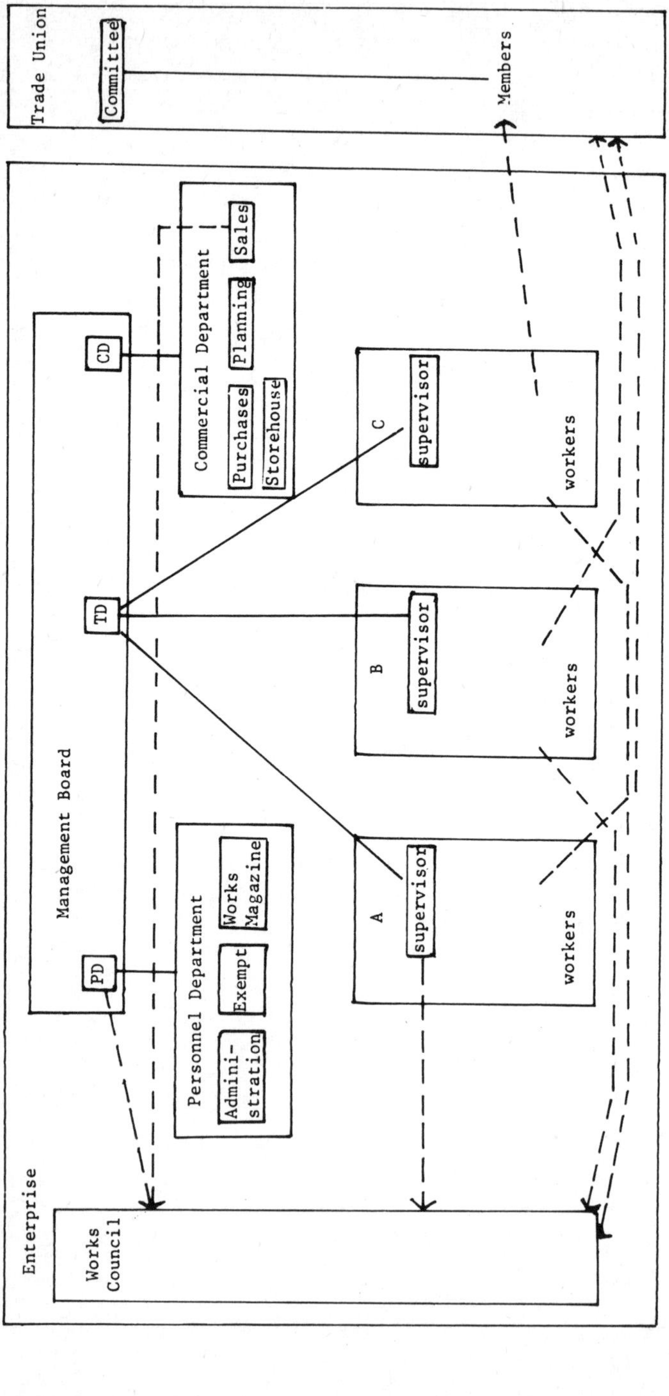


Fig. 1: Simulation of a production facility
(Organizational Chart, Initital Situation)

### 3.1a. The issue of real social interaction: psychological validity

The experiments lead to very high levels of psychological involvement, already after a short period of time, say 2 years in simulation time. This can be observed very clearly in indications like the number of drop-outs, the amount of time in which the interaction is not in terms of the simulation reality, etc. For a number of more than 2000 observations in the production-facility experiments the number of observed drop-outs is 0.6% of all participants. In about 85% of the observations the interaction is for more than 80% of the observation periods in terms of the simulation reality. The astonishing high levels of involvement are in agreement with the fact that, in the evaluative discussion on the third day of the experiment, many participants are still involved in their role and have difficulties in evaluating the events of the experiment from an outsider´s point of view.

The high level of involvement in interactions among participants also indicates that the problem of experimental demand of tradional social-psychological experiments is negligible during the simulation experiments. The ´organizational demand´ of the situation dominates it almost totally. In Van der Meer´s 24 experiments, for only two experiments the research-staff is experienced by the participants as salient for more than 10% of the observation time.

### 3.1b. Social and cognitive characteristics: structural validity

Thus our experiments clearly show that a valid social reality can be generated. However, to be relevant for social impact assessment the major social and cognitive aspects of the reference system have to be reproduced in the social reality of the experiment. We studied this problem very carefully. I will summarize our approach and results.

The model of the production facility represents a hierarchical organization with a technological, an economic and an organizational structure. We designed a broad set of criteria, characteristic for a hierarchical order of positions and reported in field studies of organizations. They cover such diverse items as the perceived distribution of influence for different fields of activities, phenomena of stress, social distances between different groups in the organization and hierarchical behavior. For 85-90% of our 28 checks per experiment (672 in total) and measured after 2 simulation years, the expected relationships, characteristic for the intended structures, can be clearly identified, both on the behavioral level and at the level of participants experiences. Moreover, by the pattern of information exchange, we carefully observed the task-stability of the different subgroups involved and their cognitive interactions. For about 80% of the 13 tests the intended characteristics are reproduced in the experiments.

Thus we may conclude that the social simulation setting has a structural isomorphy with major social and cognitive characteristics of the reference system i.c. a hierarchical production facility.

### 3.2. Reproducibility of the initial setting

The data referred to above have been measured in every replication separately. Thus it is possible to analyze the reproducibility of the experiments. Of course it needs some time for the participants to generate stable patterns of interaction. We measured the patterns after 2 simulation years, which is 2-3 hours in newtonian time. The results are very positive: on the average for only 3 of the 28 checks for structural isomorphy we do not find the intended results. This means that the simulation setting can be used for systematic experimentation and that systematic effects can be discriminated from random effects.

It should be kept in mind that these results do

not mean that the events in replicated experiments are the same. At the level of events and the ´histories´ the participants make, every exeperiment is different. Thus, replications of the initial setting at the social-structural level also disclose the potentialities of the setting for alternative histories, and in this way provide an option for generating policy alternatives. Because of the underlying structural isomorphy of the situation with the reference system, these alternatives are policy relevant.

### 3.3. Plurality of social structure and associated social realities

The research results presented so far are based on the simulation experiments of a production facility, in which all the participants are interrelated through the structural and cognitive characteristics of this facility. This structural isomorphy implies stabilized patterns of interaction between participants, e.g. formal and informal networks of communication and decision making processes. Thus the social impact of policy interventions can be studied in a psychologically and structurally valid environment. However, this does not mean that the interpretation of the situation is the same for all the participants. We discovered a positionrelated set of different experiences and interpretations of the participants, which can be interpreted very elegantly from a theoretical perspective.
For this we developed a theory on social integration, based on the concept of organizing as a form of social interaction, in which participants try to reduce equivocality and try to establish frameworks of shared meaning. Organizations and social structure in general are considered the stabilized outcome of such interaction processes. This theory can explain the position-related phenomena very convincingly. I will not elaborate on these research results here, but refer to Van der Meer´s paper and thesis 2). For this presentation it is sufficient to note that the

structure of the production facility functions for the participants as a framework for ongoing interaction and social integration through processes of stabilizing and sharing meaning.

For a large number of social situations, however, no clear-cut overall framework exists for all the participants involved. Such is for example the case for the science policy advisory councils ("sectorraden") in which persons with social backgrounds in the field of research, in the government and in organizations and groups using the result of research (e.g. industry and sometimes part of the government too) form a tripartite forum to advise on research priorities. For this situation we designed a simulation model in which some Ministeries, two research institutes (one for technology and one for social science), two production facilities and a local municipality are represented. As a primary process the planning and realization of a new district in a town called Mosdorp presents cues for the participants to interact. Moreover, the model describes a structure of interdependencies and task relations within the three societal sectors and organizations.

Bolk and Wilhelm 3) studied the emerging social structures in a series of 10 experiments. They found a manifold of social structures with distinct positions of participants and groups within these social structures, that can be interpreted from the above indicated theoretical perspective as coherent frameworks for ongoing interaction and sharing meaning.

Several groups participate in more than one of these frameworks. Thus a plurality of distinct frameworks can be generated, which also represents a plurality of interpretations of social reality and related interests. Thus social simulation experiments can generate such field-equivalent conditions in an experimental setting. The members of the science policy advisory council in these experiments are participating in different frameworks and have to handle this plurality of interpretations of the situation in their advisory work.

The research finding that the method of social

simulation can generate social plurality is very important for its potentialities as a methodology for social impact assessment. For example the problems of unexpected side-effects of policies find very often their cause in such plurality.

### 3.4. Policy interventions and strategies

We have also experimented in our models with policy interventions. In the model of the production facility three different strategies for social change have been carried out by the trade-union groups in the model. These positions were enacted by role-players. The other participants were unaware of that. In the science policy advisory council experiments, too, different advisory strategies have been enacted through members of the advisory council. Wilhelm and Bolk report on these experiments. The role-players have to realize their strategic concepts in concrete and unpredictable interactions with the other participants. Thus no prefabricated instructions can be given: the role-players act according to characteristic elements of the strategy, both on the behavioral and cognitive level. To check if the strategic concepts have been implemented, at the end of the experiments the participants are asked to charactirize the science policy advisory council´s behavior on a questionnaire, which discriminates the two strategies involved. A similar procedure has been used for the trade-union strategies in the model of the production facility.
Our findings are that in both simulation models it is possible to realize such strategies as clearly distinguishable conditions and to study their effects in the course of the experiments.

### 3.5. Cultural effects

So far, we focussed on the structural characteristics of the social reality, generated in the experiments. Of course, the attitudes and value preferences of the participants are also relevant for their capabilities to engage in stable patterns of interaction with each other. We studied this issue in our experiments with the model of the production facility. This model represents a traditional hierarchical organization. Selecting participants beforehand on more radical or conservative attitudes with respect to labour relations, we composed groups of participants with relatively radical and relatively conservative attitudes and preferences. We did several replications under each condition and found significant effects. Van der Meer reports on the results.

This example shows that the method of social simulation can be used to study the effects of cultural and structural determinants of a situation <u>independently of each other</u>. It also provides an opportunity to investigate experimentally the complex relationships between attitude and behavior in complex social settings.

### 3.6. Conclusions

Summarizing our results so far I showed that the setting of social simulation experiments can represent psychologically and structurally valid partial isomorphies of pluralistic social situations, in which political strategies can be tried out and their effects be studied systematically. Moreover, as the histories of the experiments on the level of concrete events are different for each experiment, the potentiality of the initial setting for alternative developments can be explored. Thus the method can be used as a learning strategy to explore and assess the social impact of policy interventions.

## 4. SOCIAL SIMULATION: ITS LIMITATIONS AND USE

Of course, the method also has its limitations

and shortcomings. A first point is that, although major social and structural characteristics of the reference system can be generated in the simulation setting, it always is a simplified model and one can never tell if all relevant factors have been included. A second point is that, psychologically valid as the situation may be, important factors like family relations are not included in the experimental setting. A third point is, that processes of negotiating and consulting can not be speeded up on the same scale as e.g. the production processes, and imbalances between them can arise.

The social simulation can not replace field experiments for social impact assessment. On the other hand it has characteristics which make it an interesting complementary research methodology. Compared to field experiments, its societal and organizational costs are much lower. If things go wrong in a field experiment, it is very difficult to normalize the situation and usually, the costs involved are very high. Moreover the option for systematic experimentation can lead to insights, which cannot be achieved in field experiments.

A special promising procedure arises when social simulation and field experiments are coupled. The method then becomes a kind of artificial intelligence of the setting involved. It can be used as a heuristic device to generate ideas and consciousness of the situation and, in systematic experimentation, generate insights in the future impact of policies.

One of the major problems in social impact assessment is the problem of unintended effects of policies. Part of these side-effects originate from the different interpretations of the situation and different value orientations of the societal groups involved. Also for the study of this problem the social simulation method can be helpful. The plurality of social frameworks it can generate provides an opportunity to study systematically these different definitions of social reality in real interactions.

Social simulation is often considered a somewhat exotic methodology and doing social simulation experiments certainly provides a lot of fun. However, apart from that, I hope to have shown in this article that the method has a great potentiality, both for serious research and as a strategy for learning. Within its obvious limitations, it can certainly be developed to a larger sophistication. For social impact assessment it provides a methodology which has not been used to its full extent, so far.

## REFERENCES


1. H. Guetzkow et al. (eds.): "Simulation in social and administrative science: overviews and case-examples", Englwood Cliffs (New Jersey), 1972; H.A. Becker (ed.): "Simulatie in de sociale wetenschappen" [Simulation in the social sciences], Samson, Alphen aan den Rijn, 1976; T.S. Palys: "Simulation methods and social psychology, Journal for the theory of social behaviour, 8, 1978, p.341-368.

2. F.B. van der Meer: "Organisatie als Spel: sociale simulatie als methode in onderzoek naar organiseren" [The game of organizing: Social Simulation as research method on organizational phenomena], Technological University Twente, 1983; F.B. van der Meer: "Social simulation: a research methodology and learning strategy for social impact assessment", to be published in H.A. Becker and A. Porter (eds.): "Impact Assessment Today", 1986.

3. P.W. Wilhelm and H. Bolk: "Strategy and the art of structure maintenance: on the interplay of strategy and structure in science policy", to be published in H.A. Becker and A. Porter (eds.): "Impact Assessment Today", 1986; H. Bolk, P.W. Wilhelm: "Simulatie van sektorraden: model en validiteit" [Simulation of sectorcouncils: model and validity], Staatsuitgeverij, Den Haag, 1984.

## SOCIAL IMPACT ANALYSIS OF FOUR ENERGY SCENARIOS

H.P. Peters*

### ABSTRACT

The program group "Technology and Society" of the Nuclear Research Centre Jülich was assigned the task by the Federal Ministery of Research and Technology (FRG) to investigate the social impacts (Sozialverträglichkeit) of four energy scenarios developed by a Parliamentary Enquete Commission consisting of experts and politicians. The Julich study group based their investigation on a complex model linking the technique of value tree analysis with traditional methods of technology assessment and the incorporation of participatory involvement by citizens.

First, by interviewing representatives of stakeholder groups the concerns and criteria relevant to the energy debate were collected and sorted, secondly, the impacts of each dimension were measured or estimated (via expert ratings or judgements) and thirdly, the weights and evaluations for each criterion and option were assigned by so called planning cells. Planning cells consist of groups of citizens who are selected by random process and are given paid leave from their workday obligations for a limited period of time to work out solutions for social problems. A total of 24 planning cells were organized throughout Germany to evaluate the four energy scenarios and to formulate recommendations for the policy maker. One outstanding result was that most citizens favored the more moderate scenarios, but were almost equally devided in their preference distribution with respect to the pronuclear and non-nuclear option. The paper describes the basic methodology, gives a review of the results and discusses, the pitfalls and merits of this three-fold approach for policy making.

KEY WORDS

Technology and Society, social impacts, energy scenarios, value tree analysis, participatory involvement, expert ratings, planning cells

---

*Nuclear Research Centre Jülich, Program Group Technology and Society, P.O.Box 1913, D-5170 Jülich

## 1. Introduction

The opposition of parts of the German society against nuclear energy has demonstrated that energy supply is not longer seen as a technical question alone, which can be left to experts. Problems with the acceptance of nuclear power, the perceived limits of fossil energy resources, the increase of oil prices and environmental damages caused by energy systems made the energy problem to a main topic for the political system and public discussion. The German parliament reacted to that development with the institution of a so called "enquete commission", which started its work in March 1979 and finished its report in June 1980 (Deutscher Bundestag 1980).

Among others the commission developed four scenarios of the energy system in the FRG until the year 2030. Those scenarios sketch different "paths" of development of the existing energy system in the future. All scenarios are constructed according to similar boundary conditions (growth of population, economic growth) and allow the same energy related services. The way in which the energy services (light, power, heat) are provided differs between the four paths (see fig. 1). Path 1 will use fossil and nuclear energy sources to a greater extend than today to meet the increased demand for energy services. Path 4 as the other extreme will make great efforts for energy conservation to reduce the consumption of fossil energy sources, eliminates nuclear energy and secures energy supply mainly with renewable energy sources (solar energy, wind, water, biomass etc.) and a small amount of fossil energy sources. Path 2 en 3 are located between the two extremes. They use the same amount of fossil energy sources. Path 2 in addition uses nuclear energy while path 3 substitutes this contribution by greater efforts to conserve energy.

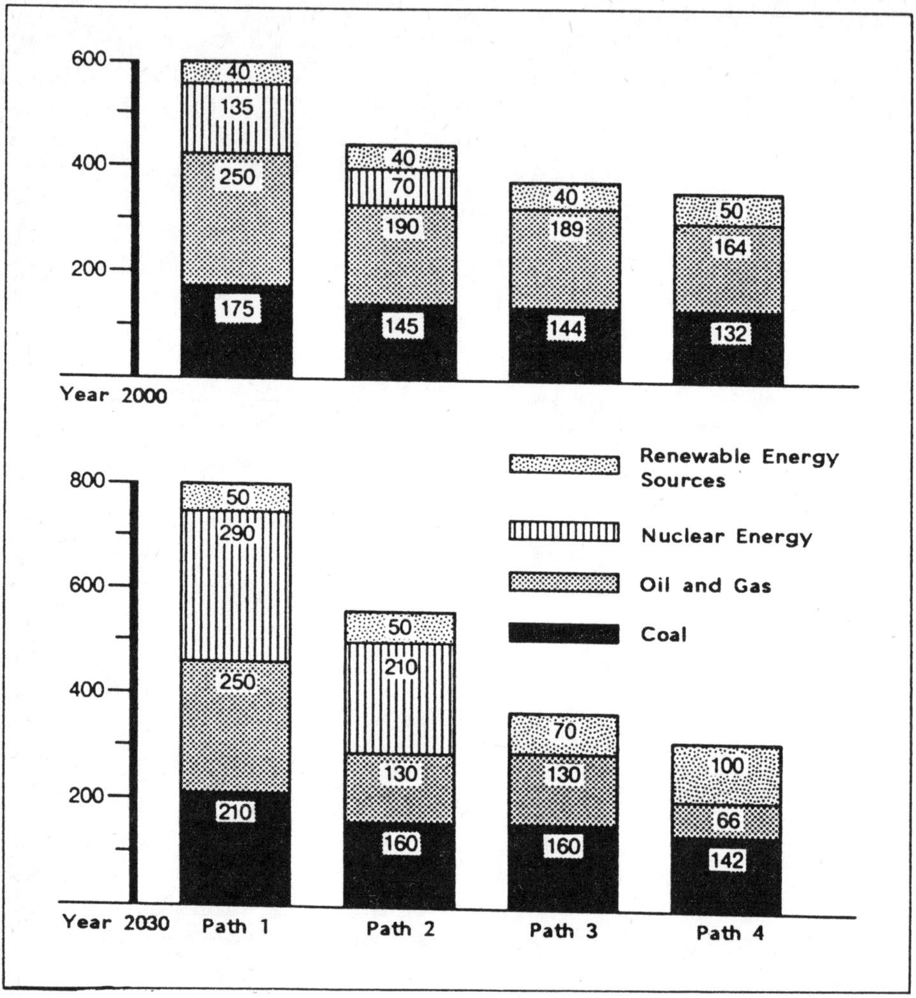


Figure 1. Primary energy supply of the four paths by different energy sources (in Mio t coal per year)

The enquete commission agreed to a set of four criteria to be used for the evaluation of energy systems:

1. economy,
2. international compatibility,
3. environmental compatibility,
4. social compatibility.

The first three criteria do not need further explanation, but the fourth criterion, the "social compatibility", is a new aspect in the evaluation of energy systems and needs further explication.

Social compatibility is a somewhat vague label, that has to be operationalized if used in a scientific study. Social compatibility in our study is understood as the correspondence of a certain development (e.g. in technology, economic order or - in our case - energy policy) to the social value system. We assume that the acceptance of an energy system depends on the amount of perceived value violation. To predict value violations a social impact assessment had to be conducted.

To analyse social compatibility in the described meaning it is necessary to comprehend the social values in a first step and then to determine the points of value violation by the discussed alternatives in a further step. The four paths of the enquete commission were selected as alternatives to be evaluated in our project.

Our project consists of three steps, each being dependent on the preceding steps:

1. The social criteria used to evaluate energy systems had to be determined. For this purpose we used the instrument of "value tree analysis" (Keeney et.al. 1984).
2. An impact analysis was conducted to describe the consequences of each path on the criteria found in the first step.

3. Randomly selected citizens in so called "planning cells" (Dienel 1978) were confronted with the results of the impact analysis and their perception of the advantages and disadvantages, their values with respect to energy problems and their preferences for one of the four paths were collected.

From the results of the planning cells we conclude on the factors determining the "social compatibility".

## 2. Value Tree Analysis

An analysis of social goals with respect to energy politics has to deal with the problems of value pluralism, which means that different people or groups aim at different goals, and the overcharge of "normal citizens" if directly asked for their value orientation in energy politics. The value orientation influences people implicitly, so that the values cannot be directly expressed by people. Looking for social goals one can concentrate on interest groups which try to realize a few values in a professional way and are able to formulate their goals explicitly. If one includes not only one group but as many interest groups as are engaged in public discussion about a certain problem (e.g. energy supply), one can be sufficiently sure to get a rather complete catalog of relevant social goals.

We used the value tree analysis as an instrument to gather the values of organized social groups. Nine social groups, starting with the organization of the German industry and ending with a German nature protection organisation, participated in this value tree analysis.

In conducting a value tree analysis, one tries to gather the values of interest groups during interviews with representatives of those groups and to arrange the expressed values in a hierarchical structure. The result is called a "value tree". The general values are located near the stem while the more specific values are positioned at the cusps of the branches.

For each of the interviewed nine groups a seperate value tree was constructed. The interest groups were asked whether they could agree to that value tree. In a second step the nine seperate value trees were combined to a common one. This was possible because the general structure of the different value trees was rather similar. They differed in the branches where the value structure was differentiated very deeply into sub-values and sub-sub-values. E.g. the representatives of the industry split the general value of economy into a lot of sub-values while the representatives of the environmental protection organization differentiated the value of a healthy environment to a high degree.

The common value tree was used to construct a system of criteria, which was applied in the following steps as a systematic for the impact analysis and the collection of citizens' preferences and evaluations in the planning cells. It was necessary to simplify the structure of the value tree. The basic structure was taken as the main criteria but the complex detail structure had to be reduced to few sub-criteria which were further operationalized by indicators.

The system of criteria includes the following eight criteria:

1. Financial and Material Requirements,
2. Security of Supply,
3. Economic Effects,
4. Environmental Impacts,
5. Health and Safety,
6. Social Impacts,
7. Political Impacts,
8. International Effects.

Each of the eight criteria includes several sub-criteria which are split into indicators. The system of criteria is nothing else but a broad and ordered list of relevant values with respect to energy systems.

## 3. Impact Analysis

The impact analysis of the four energy paths raises severe methodological and practical problems because the purpose is not to evaluate real (existing) systems but to evaluate scenarios. Logically all statements of the impact analysis are conditional predictions: what would the world look like if an energy system like path 1, 2, 3 of 4 would be implemented - all boundary conditions assumed constant.

On one hand our study is designed for the purpose of policy consultation, on the other hand a scientifically satisfactiory impact analysis was possible only for few indicators. But political decisions have to be made today - under uncertainty. So a pragmatic approach had to be chosen. Our impact analysis aimed to summarize all information which should be considered in energy policy decision making and to structure this information according to our system of criteria. Experts with different points of view participated in our impact analysis. We tried to make clear on which topics they agreed and where they disagreed.

Information about the impacts of different energy systems was collected by analysis of expert literature, participation of energy experts of different points of view, and sub-contracts with institutions which have a high reputation in the field of impact analysis by their former research. This technology assessment was not aimed to clarify few special problems but to summarize the existing but scattered knowledge and organize it in a way that makes it useful for decision making (Albrecht/Stegelmann 1984).

## 4. Planning Cells

Twenty four planning cells were run from June 1982 to March 1983 in 7 German towns which dealt with the "social compatibility of energy supply systems". Those planning cells were organized by the Research Institute for Citizens' Participation and Planning Methods at the University of Wuppertal; the programme of the planning cells was developed by the research institute of Wuppertal and our research team at Jülich.

A planning cell is a group of about 25 randomly selected citizens, who come together for a limited period (about one week) to deal with a given task, e.g. to express an opinion on a certain planning problem. The participants of the planning cells are paid for this job. This procedure has been developed by P.C. Dienel to contribute to the solution of two social problems: the demand for more and new forms of citizens' participation and the demand for good planning (Dienel 1978).

According to these intentions, planning cells were formerly used as participation methods and instruments for policy consultation. In our project the planning cell is used as an survey instrument in addition to these functions. The participation approach was not given up in our project but was modified in the way that the informed citizens were asked to express normative decisions to be used for the consultation of political decision makers. The main interest was not on the facts (description of objective consequences) but on the values of preferences. A "citizens' report" summarizes the recommendations of the participants of the planning cells for political decision makers (Dienel/Garbe 1985). Participation of citizens in the highly complex field of national energy policy is hardly possible in relation with technical details, but it is very useful to state the value-oriented goals of national energy policy and to evaluate the ways in which the existing energy system can develop.

The participants of the planning cells were informed about the results of the impact analysis by experts, video-films, and written material. During the few days of the planning cell the participants had to evaluate the four energy paths on the eight criteria and to weight the criteria according to the ascribed relevance for decision-making in the energy field. Furthermore the attitudes to and opinions about some policy strategies and technologies were collected and sociodemographic data recorded.

During the planning cells the participants had to choose one out of the four paths as their first preference. The result of this decision process was that the more moderate paths 2 and 3 received the greatest support: 41,5% choose path 2 and 39.4% path 3 as their first preference. A relevant minority of 16.0% choose path 4 while path 1 was preferred only by a very small fraction of 3.1%.

The results of the planning cells cannot be extrapolated to the population of the FRG. This is prohibited by three reasons:

1. Although the people who were invited to join a planning cell were randomly selected only a fraction of about 20% followed that invitation. Some subgroups within the population were over-represented, others were under-represented.
2. During the four days of the planning cells the participating citizens had an intensive exchange of ideas, opinions, and judgements with energy experts and among themselves. With respect to their information level they differed from the general public to a great extend.
3. The participants of the planning cells acted according to a special role, that of a "citizens' expert", which differes from round-day-roles and e.g. implies a duty to consider the common welfare. This certaincy has changed the perspective in which the problems were seen and has probably influenced the decisions.

The behaviour of decision making observed in the

planning cells is not comparable to that in other situations, e.g. political elections, but for policy consultance purposes this "artefact" is particularly meaningful, because the policy relevant preferences of citizens are expressed therein more clearly and less distorted by partial interests and missing or wrong information than in other procedures of data collection. Besides, the expression of opinions under the pressure, to have to be justify one´s own view against contradicting arguments of other participants, leads to more reliable and consistent results than e.g. in public opinion research. The decisions of the citizens in our planning cells cannot be discarded as expressions of naive and superficial opinions by uninformed citizens, but have to be considered as judgements that were made after an extensive discussion of the problem.

In order to obtain the value preferences of the planning cell participants they were asked to give weights to the eight criteria which had to add to 100. The criteria with the highest average weights are "Health and Safety", "Environmental Impacts" and "Security of Supply" (see fig.2). The criteria "Economic Effects", "Financial and Material Requirements" and "Social Impacts" are of middle importance. As least important for the evaluation of energy systems the criteria "Political Impacts" and "International Impacts" were rated. As difference between the weights of the criteria 4-6 (Environmental Impacts, Health and Safety, Social Impacts) and the weights of the criteria 1-3 (Financial and Material Requirements, Security of Supply, Economic Effects) an index was constructed. This index describes the dominance of postmaterialistic or materialistic values. It correlates significantly with the path preference: participants who preferred path 1 are likely to be oriented towards "materialistic" values; people who gave paths 3 or 4 the first preference tend to emphasize "postmaterialistic" values. Participants who decided in favour of path 2 cannot be clearly classified into one or the other group.

A content analysis of the verbal explanations of the selection of one of the four paths proves that the attitudes towards several energy sources is closely related to the path preference. It is not surprising that people who were in favour of paths 1 or 2 perceive nuclear energy positive and those who preferred paths 3 or 4 had strong objections against its use. This result corresponds to the observed correlation of a nuclear energy attitude-scale with the path preference. The difference in the evaluation of fossil energy sources are not so strong but respondents in favour of paths 1 or 2 more often argue by referring to negative consequences of fossil energy sources than people who preferred paths 3 or 4. The path 3/4 preference group more strongly emphasizes the advantages of renewable energy sources than the path 1/2 preference group. With respect to energy conservation all groups mentioned only positive arguments - without any differentiation between the proponents of path 2 and those of path 3. This result is confirmed by the missing correlation of an index "desired political influence on energy conservation" with the path preference. Regardless of the path preference nearly all participants of the planning cells agree to the necessity of political efforts to foster energy conservation.
The participants of the planning cells had to evaluate the four paths on the eight criteria as one of their main tasks. The paths 3 and 4 which do not use nuclear energy on an average were evaluated more positively than the paths 1 and 2 on the criteria "Environmental Impacts" and "Health and Safety". The extreme paths 1 and 4 were evaluated more negatively than the moderate paths 2 and 3 on the three economy-related criteria (see fig. 3). The proponents of a path of course evaluate this path more positively than those who are in favour of another path.

Conflicts between goals are more often observed among the proponents of path 2 compared to those of path 3. About 86% of the proponents of path 2 evaluate path 3 in some aspects (environment and health/safety) more positively than the path 2. That means that they perceive other alternatives superior to their own preference in some dimensions. On the other side the portion of the path 3 proponents who experienced conflicts among goals is only 51%. In addition, the average strength of conflicts is greater in the path 2 preference group than in the path 3 preference group. This leads to the conclusion that path 2 proponents would more easily accept path 3 than path 3 proponents would accept path 2 - assumed that both paths can technically and economically be implemented and provide the expected comfort level.

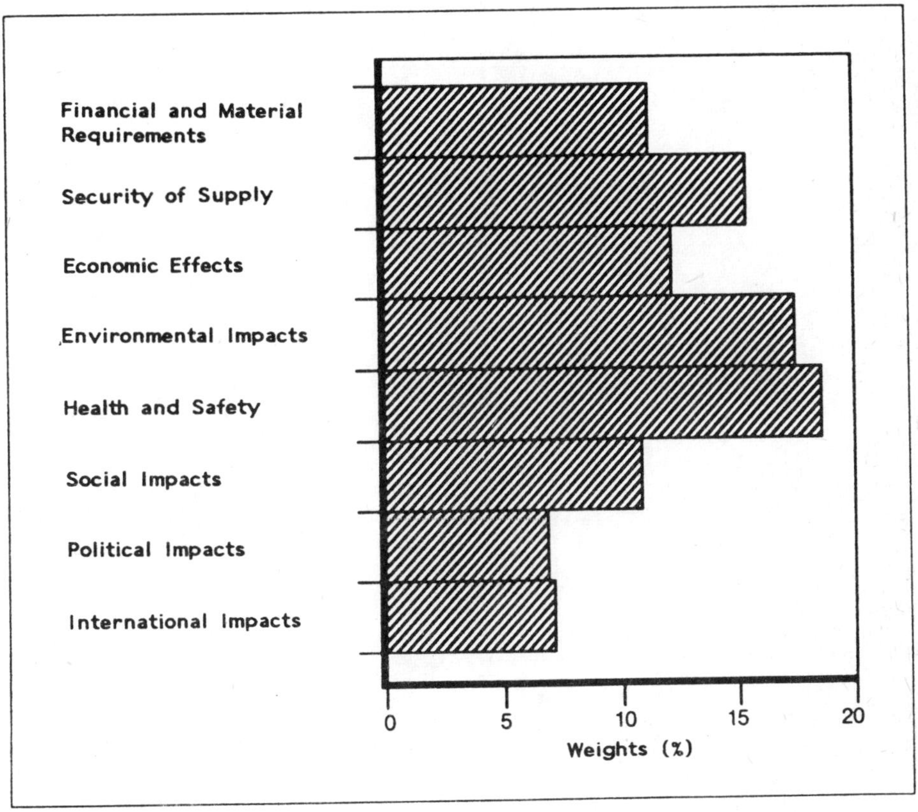


Figure 2. Weights given to eight criteria by the participants of the planning cells (means).

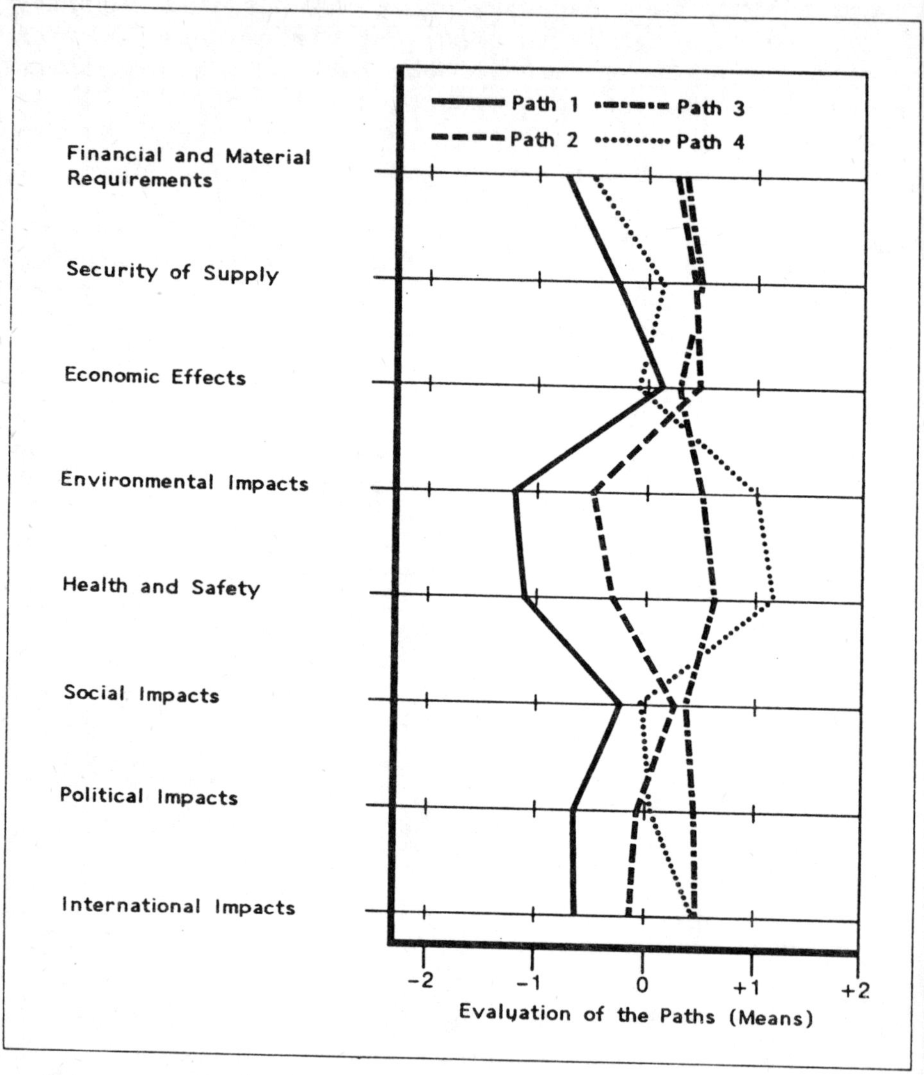


Figure 3. Evaluation of the four energy paths according to the judgements of the participants in the planning cells (means).

A simple decision model (Multi-Attributive-Utility-Model) was used to predict the expressed path preference for every participant by the criteria weights and perceived consequences of the four alternatives (Keeney/Raiffa 1976). The model works in the following way:

1. For each alternative the sum of all evaluations multiplied with the corresponding criteria weights is calculated.
2. The resulting sums are compared; the alternative with the highest score is called "predicted path preference".

For every participant the model gives the mean of the evaluations on the eight criteria weighted according to the ascribed relevance of the criteria. If a person strongly emphasizes "environment" and "health/safety" the evaluations on those criteria influence on the resulting score particularly strong. That leads to a preference towards paths 3 and 4. If a person gives highest weights to the economic criteria, it is more likely that his predicted preference will be path 2 or 3.

To check the explicative power of the model, its "predictions" were cross-tabulated with the expressed preferences. In about 50% of all cases the model predicts the correct path. That shows that the individual decisions are connected to the perceived consequences on the value dimensions. With four alternatives, however, the probability of a correct prediction by chance is already 25%. The "predictive power" of the model therefore is not very high.

More interesting than the limited predictive power of the model are the observed systematic deviations from the correct predictions. The results of the model are strongly biased in two ways (see fig. 4):

1. The model predicts more often the extreme paths 1 and 4 as were expressed by the citizens themselves.

2. The results of the model are systematically shifted toward the paths 3 and 4.

Additional factors, which are not covered by the model, influence the decisions of the citizens and shift the preference in the direction of path 2. Two of those factors could be identified by the content analysis of the verbal explanations:

1. The criterion of "Security of Supply" seems to have a higher weight intuitively than is expressed in direct rating. Arguments concerning the security of supply were most often used - more often than even health/safety or environment - to legitimize the decision. Compared with path 4 path 3 is evaluated more positively on the criterion of security of supply.
2. A lot of participants of the planning cells doubt whether paths 1 and 4 can be efficiently implemented. The MAU-Model however assumes that all alternatives are realistic and provide the same comfort level.

Other possible factors influencing the expressed preference, which are not covered by the model, could be the general tendency to avoid extremes or the orientation on reference groups and authorities.

## 5. Political implications

The participants of the planning cells were not a representative sample of the whole population in the FRG because of the non-representative selection of the towns in which planning cells were conducted and the high proportion of citizens who refused to participate. Furthermore the participants are distinguished from the general public by their high level of knowlegde and their role as "citizen experts". Statistical inferences of the results in the planning cells to the total population should not be made. We obtained, however, insights into the correspondences of value preferences, perception of consequences, and attitudes, which will probably be true for the total population and which lead to politically relevant implications.

First it is obvious that extreme alternatives are favoured only by minorities. The larger proportion of the participants was in favour of a moderate policy that tries to find the best solutions for defined problems but does not use energy policy as an instrument for social changes.

A large proportion of the citizens was afraid by nuclear energy. This fear cannot be reduced by information alone. A broad social consent about the use of nuclear energy seems to be unlikely now; an acceptance can be expected if - after the nuclear fuel cycle has been closed - the number of symbolic targets decreases and the nuclear energy controversy is displaced by other political topics.

| | expressed preference | predicted preference |
|---|---|---|
| Path 1 | 3.5% | 6.4% |
| Path 2 | 49.4% | 24.5% |
| Path 3 | 38.5% | 31.5% |
| Path 4 | 8.6% | 37.6% |
| | 100.0% (N=314) | 100.0% (N=314) |

Figure 4. Predicted versus expressed preferences of the participants of the planning cells for the four energy paths: The predicted preferences are calculated according to a decision model. The expressed preferences were gathered by a questionaire.

Measures to conserve energy are favoured by nearly all participants. For that reason there are good conditions for a policy that promotes energy conservation. It is still an open question how the population would react if serious negative consequences of an energy conservation policy would occur or the expectations in that policy would not be realized.

An unambiguous answer to the question "Which path is the most socially compatible one?" cannot be given by our study. The extreme paths 1 and 4 are certainly not socially compatible. Their realization would hurt central values of a relevant group within the population. The realization of each of the paths 2 or 3 is connected with specific risks for the social compatibility which have to be balanced by the political process. The social compatibility of path 3 presupposes that it realizes the economic promises in the eyes of the path 1/2 preference group for which those values are very important. The social compatibility of path 2 on the other hand is to be expected only if the problems with the acceptance of nuclear energy can be overcome or this controversy looses its relevance.

## LITERATURE

Albrecht, G./Stegelmann, H.U. (eds.), 1984: Energie im Brennpunkt. Zwischenbilanz der Energiedebatte. München: High Tech

Deutscher Bundestag, Presse- und Informationszentrum (ed.), 1980: Zukünftige Kernenergie-Politik. Kriterien - Möglichkeiten - Empfehlungen. Bericht der Enquete-Kommission des Deutschen Bundestages. Teil I und II. Bonn

Dienel, P.C., 1978: Die Planungszelle. Eine Alternative zur Establishment-Demokratie. Opladen: Westdeutscher Verlag

Dienel, P.C./Garbe, D. (eds.), 1985: Zukünftige Energiepolitik. Ein Bürgergutachten. München: High Tech

Keeney, R.L./Raiffa, H. (1976): Decisions with Multiple Objectives. Preferences and Value Tradeoffs. New York: Wiley

Keeney, R.L./Renn, O./von Winterfeldt, D./Kotte, U. (1984): Die Wertbaumanalyse. Entscheidungshilfe für die Politik. München: High Tech

Renn, O./Albrecht, G./Kotte, U./Peters, H.P./Stegelmann, H.U. (1985): Sozialverträgliche Energiepolitik. Ein Gutachten für die Bundesregierung. München: High Tech

SOCIAL IMPACT ASSESSMENT IN NATURAL RESOURCE DECISION MAKING: TOWARD A STRUCTURAL PARADIGM

Dr. Robert M. Muth*
Dr. Robert G. Lee**

ABSTRACT

As social impact assessment continues to mature as an area of professional endeavor, social scientists are becoming increasingly aware of the importance of social structure as a variable requiring analysis in social impact assessment studies. This paper reports the results of research in which community social structure was specifically examined as an independent variable in a regional system of 52 rural communities in Alaska. A macrostructural survey instrument was employed to inventory elements of community social structure provided by key informant interviews and a variety of secondary sources. Communities were then ranked according to their relative structural differentiation by means of Guttman scale analysis. Multiple regression analysis indicated that structural differentiation possesses greater predictive power than traditional aggregate variables (population level, rate of population increase) in explaining rates of community social pathology (mental health rates and divorce). The utility of the methodology for conducting social impact assessment is discussed. Suggestions for further research are presented.

KEY WORDS

Natural resource decision making, community social structure, macrostructural survey instrument

---

* Pacific Northwest Forest and Range Experiment Station, U.S.D.A. Forest Service, Juneau, Alaska, U.S.A.

** College of Forest Resources, University of Washington, Seattle, Washington, U.S.A.

## 1. Introduction

Throughout the history of the United States, natural resource development and community change have often occurred simultaneously. This process has resulted in social strains and stresses of varying magnitudes and it has occasionally manifested itself in a variety of forms of social disorganization. In response to passage of the National Environmental Policy Act of 1969, the systematic identification of social effects resulting from federally-sponsored natural resources development projects has become an integral part of the environmental assessment process.

As social impact assessment became institutionalized as an area of professional endeavor, it came under increasing scrutiny from a number of critics (Wilkinson, et al, 1980; Wilkinson et al., 1982; Muth, 1983; Wilkinson et al., 1984; Lee et al. 1984; and Summers and Branch, 1984). If one views the social impact assessment literature in its most comprehensive form (applied research, textbooks, professional newsletters, agency manuals, consulting firm reports, administrative studies, and environmental impact statements), a number of critiques can be offered.

First, social impact assessment has been atheoretical. Second, social impact assessment has generally been descriptive rather than explanatory. Third, social impact assessment has often ignored the historical context of social change. Fourth, social impact assessment has focused on cases where social disorganization is manifest and ignored communities that are coping successfully with change. Finally, and perhaps most important for us as social scientists, social impact assessment has generally failed to include as a variable for analysis that constellation of social constructions which pattern human behaviour referred to as institutional structure. This omission has occurred despite the fact that social scientists in other subdisciplines continue "... to converge on the viewpoint that social change is essentially structural change or change of social structure itself." (Kim, 1973, p. 463).

Social impact assessment has usually concentrated on aggregated variables such as population size, population increase, income, and perceptions and attitudes, rather than on the structural variables that characterize the nature of any social system. Structural analysis has the potential to enhance greatly our understanding of the relationship between natural resource development and social change, precisely because "... it is through this type of analysis that we are able to see the social basis of a great deal of behavior and thought that conventional thought once ascribed to biological, psychological and even supernatural forces." (Nisbet, 1974, p. 74).

This paper suggests a structural approach for improving explanatory and predictive capabilities for social impact assessment in natural resource planning and decision making. Results are presented from an exploratory application of a macrostructural inventory method to resource extraction communities in Alaska.

## 2. A structural approach to social impact assessment

### (a) Importance of Institutional Structure.

We use the term "social structure" to refer to the persistent patterns of relationships between elements of a society. Those elements are specified in terms of social institutions - organized and enduring patterns of social behavior that embody social values or other specialised meanings necessary for the functioning of a society.

The relevance of social structure for analysis in social impact assessment research is derived from the theoretical proposition that anomie and social pathology result when the rapid division of labour (i.e., differentiation in the complexity or specialization of institutional structure) outstrips concomitant development of the integrating mechanisms of society. Referred to by Durkheim (1933 [1893]) as "the anomic division of labour" newly-differentiated functions become isolated and disconnected from society because the overarching normative structure fails to envelope them within the existing moral order. Although tentative empirical verification of this theoretical position has been obtained in studies examining the relationship between the division of labour and crime rates (Webb, 1972; Krohn, 1978) and between the division of labour and suicide (Miley and Micklin, 1972), further research is necessary to determine the applicability of structural variables to specific social impact studies.

A large number of recent social impact studies has concluded that the wide range of social stresses and pathologies (such as drug and alcohol abuse, homicide, divorce, mental health admissions, crime, suicide, child abuse, domestic violence) experienced by rural communities is attributable to dramatic increases in population levels associated with resource development (see the review by Summers and Branch, 1984). Adjustment problems are said to be exacerbated by the additional pressures placed on pre-existing levels of infrastructure, human services, and economic opportunities (Davenport and Davenport, 1979, 1980) or by cultural conflicts (Geisler, et al, 1982; De Kadt, 1979). The theoretical view that structural variables may be more important determinants of social pathology than population change inspired us to examine the relationship between natural resource development, institutional structure, and social pathology in Alaskan resource-based communities.

(b) Setting.

The State of Alaska is in rapid transition. The State's economy is heavily dependent on petroleum and mineral development, timber harvest and agriculture, commercial fishing, government, and most recently, tourism. A recent impetus to the development of Alaska's natural resources is the increasing availability of developable lands conveyed by the federal government to the State, Native corporations and private individuals through the Alaska National Interest Lands Conservation Act of 1980. A second major stimulus influencing resource development in Alaska is the fact that State government anticipates several decades of revenue from petroleum leases and severance taxes estimated to be in the tens of billions of dollars. The State legislature has committed millions of dollars to capital-improvement budgets, low-interest loans and other programs designed to spur economic development of the State's natural resources. These events have resulted in increasing levels of structural differentiation in most Alaskan communities. Also, during this period, Alaska's population increased from 300,382 in 1970 to 401,851 in 1980 - an increase of approximately 33 percent.

Fifty-two communities, comprising the universe of rural communities in southeast and southcentral Alaska were selected for case-study to explore the relationship between community institutional structure, population increase, and social pathology. This region of Alaska is the primary location for public and private timber harvesting and supplies a significant share of the fishing, mining and tourism revenues as well.

(c) Research design and analysis.

One limitation of using institutional structure as a variable in social impact assessment research has been the lack of suitable methodological approaches and analytical techniques for quantifying and statistically manipulating structural data. We selected a methodological approach for inventorying institutional structure that had been successfully tested and replicated in numerous studies world-wide (Young and Young, 1973), but that had not yet been applied to social impact assessment. With certain modifications, the collection of data on the structure of community institutions was guided by Young (1977).

The Young methodology relies heavily on the visible, concrete components of community structure that can be publicly discriminated. The macrostructural survey was constructed to collect data in such a way as to identify the presence or absence of individual institutional items in a community. The survey form contained a checklist of institutional items such as a grocery store, restaurant, paved streets, a telephone system, and a hospital.

After a macrostructural inventory had been conducted for each of the 52 communities, Guttman Scale analysis (Guttman, 1950) was applied to the institutional data. Seven sectoral subscales were constructed. A general institutional differentiation scale was developed from the institutional items forming the sectoral subscales (see Figure 1).

Figure 1. General Structural Differentiation Scale(*)

| Scale Step | Structural Item | Percent Discrimination | Errors |
|---|---|---|---|
| 1 | Elementary school | 96 | 4 |
| 2 | Vehicle Fuel Retailer | 94 | 5 |
| 3 | Electrical System | 92 | 7 |
| 4 | Restaurant | 86 | 6 |
| 5 | Telephone Network | 84 | 3 |
| 6 | Hotel | 78 | 6 |
| 7 | Bar (Tavern or Lounge) | 74 | 10 |
| 8 | Gift shop | 66 | 8 |
| 9 | Grocery Store | 62 | 7 |
| 10 | Ambulance | 54 | 7 |
| 11 | Specialty Contractor | 52 | 11 |
| 12 | Bookkeeper | 48 | 7 |
| 13 | Sewage Treatment Plant | 46 | 8 |
| 14 | Air Taxi Service | 44 | 4 |
| 15 | Paved Roads | 42 | 5 |
| 16 | X-ray machine and/or Tech | 40 | 4 |
| | Bank | | 3 |
| 17 | Public Airport | 36 | 1 |
| | Hardware Store | | 6 |
| 18 | General Practitioner (MD) | 34 | 4 |
| | Clothing Store | | 4 |
| 19 | Newspaper | 32 | 1 |
| | Dentist | | 2 |

| Scale Step | Structural Item | Percent Discrimination | Errors |
|---|---|---|---|
| 20 | Travel Agent | 30 | 1 |
| | Insurance Agent | | 1 |
| 21 | Lawyer | 28 | 0 |
| 22 | Dry Cleaner | 26 | 3 |
| 23 | Junior High School | 24 | 4 |
| | Psychologist | | 2 |
| 24 | Hospital | 22 | 5 |
| | Mental Health Clinic | | 2 |
| 25 | Veterinarian | 20 | 4 |
| 26 | Community College | 16 | 3 |
| | Mortuary | | 4 |
| 27 | Pet Store | 12 | 4 |
| 28 | Landscape Architect | 10 | 2 |

Coefficient of Reproducibility = .91
Coefficient of Scalability = .69
(*) Produced by Guttman Scale analysis of macro-structural survey data on 52 communities in southeast and southcentral Alaska.

Using the ranking of communities generated by the scalogram analysis as an operational definition of structural differentiation, multiple regression analyses were performed to determine the strength of association between structural differentiation (and other, potentially-confounding independent variables) and measures of social pathology.

Social pathology was operationally defined as community rates of violent deaths, mental health admissions, and divorce. These three measures were selected as dependent variables because of their reliability in the Alaskan context. (For a discussion of data selection and analysis, see Muth, 1985). Six independent variables (structural differentiation, 1980 population, rate of population change from 1970-1980, Native/non-Native population composition, average annual precipitation, and unemployment) were entered, stepwise, into regression equations for each dependent variable.

(d) Results

Tables 1 and 2 present the results of step-wise multiple regression analyses. They confirm the existence of a relationship between structural differentiation and mental health admissions ($R=.488$, $p < .001$) and between structural differentiation and divorce ($R=.340$, $p = .029$). No statistically significant relationship was obtained for differentiation and violent deaths (1).

Table 1. Stepwise regression analysis of divorce for communities in Southeast and Southcentral Alaska (differentiation)

| | Statistics for individual variables | | | |
|---|---|---|---|---|
| Variable | B | F | Significance | Simple R |
| Differentiation | 2.470 | 5.04 | .029 | .340 |
| Precipitation | -.334 | 1.89 | .227 | -.174 |
| Percent Native | -.312 | .58 | .458 | -.229 |
| Unemployment | -.742 | .23 | .632 | -.035 |
| (Constant) | 69.204 | 4.96 | .031 | -- |

Statistics for the Equation as a whole:

Multiple R = .404 $R^2$ = .16 F = 2.29 p = .07

Table 2. Stepwise regression analysis of mental health admissions for communities in Southeast and Southcentral Alaska (Differentiation)

| | Statistics for individual variables | | | |
|---|---|---|---|---|
| Variable | B | F | Significance | Simple R |
| Differentiation | 34.372 | 14.91 | > .001 | .488 |
| Precipitation | -5.473 | 6.14 | .017 | -.299 |
| Percent Native | -1.449 | .18 | .669 | -.252 |
| Unemployment | -.031 | .06 | .801 | .019 |
| (Constant) | 517.013 | 4.23 | .045 | -- |

Statistics for the Equation as a whole:

Multiple R = .582 $R^2$ = .338 F = 6.025 p = < .001

e) Discussion

Unfortunately, social science research often suffers from data limitations that make unambiguous interpretations difficult. This study is no exception. But despite the limitations of the data, a number of interpretations can be drawn about the relationship between structural differentiation and social pathology.

Significantly, the data on mental health admissions represents the "cleanest" of the social pathology data. People admitted to either mental health clinics or the Alaska Psychiatric Institute were categorized according to their community of residence. Of the three categories of social pathology employed in this study, mental health admissions most accurately reflect the community social structure which "produced" them. As a consequence, the relationship obtained between structural differentiation and mental health admissions should not be dismissed casually.

One possible explanation for this association is that communities characterized by highly-complex social structures are more likely to possess the specialized knowledge capable of defining and treating mental illness. Communities with a psychologist or mental health clinic contain the diagnostic and referral skills necessary to identify symptoms of mental illness as they are defined professionally. Individuals with mental disorders who live in less-differentiated villages may not be identified as such; their care and treatment may reside with more traditional structures such as the extended family or church. Because they live in communities lacking the specialized expertise capable of "defining" their behavioural problems as mental illness, they may escape being "counted" in the State statistics on mental health admissions. In this case, the incidence of mental health disorders may be more an artifact of the presence of specialized institutional components capable of defining them as such rather than any causal association with higher structural differentiation.

An alternative interpretation of these findings, however, lends support to the Durkheimian notion of the anomic division of labour - a case in which the structural differentiation of modernity occurs at such a rapid pace that specialized structural components lie outside the normative umbrella of society and produce a state of anomie and related social pathologies.

One important finding which supports this interpretation regarding structural influences on social pathology is the fact that social pathology exhibited no dependence whatsoever on rate of population increase, a variable to which many negative social impacts are attributed in much of the social impact assessment literature. This suggests that social pathologies may be related to social phenomena other than rate of population increase. One would expect this to be true especially in Alaska, where seasonal population increases (and declines) occur frequently enough that adaptive institutional mechanisms for mitigating the potentially-disruptive effects of population fluctuations may have been incorporated into the existing social structure over time.

Perhaps "boomtown" type situations evidence high rates of social pathology because their institutional structures differentiate at a rate so rapid that the existing normative order becomes outmoded and incapable of carrying out its regulation and integration functions, rather than because of dramatic population increases. The existing social (especially normative) structure may not contain adaptive capacity (including solidarity-building voluntary associations and other mediating structures) sufficient to integrate structurally-differentiated components into the normative "fold". Under these conditions, structural differentiation contributes to anomie and social disintegration instead of community cohesiveness.

## 3. Conclusions

The above interpretation must be regarded as tentative because this study did not systematically measure the <u>rate of increase</u> in structural differentiation. Rather, increases in structural differentiation were assumed to be occurring in response to the increases in economic development. Since growth and decline in population may take place at rates independent of the rate of structural differentiation, neither population level nor rate of population increase can serve as proxy measures of structural differentiation.

In retrospect, the rate of increase in structural differentiation is a crucial variable. This study would have benefited greatly from a quasi-experimental research design as called for by Young and Young (1973), in which historical baselines of structural differentiation were obtained for prior five- or ten-year intervals through key informant interviews. Comparison of previous baseline periods with the current status of structural differentiation would allow computation of the rate of increase (or decrease) in institutional structure.

Future research of this nature must incorporate measures of the value structure and normative structure in addition to relying on the publicly-discriminated visible structures emphasized here. Values and norms serve to pattern human behaviour just as surely as do roles and artifacts. Future research could improve upon the contributions of structural sociology in social impact assessment studies by employing survey instruments and participant observation to ascertain the influence of the value and normative dimensions of institutional structure on social pathology.

The results of the Alaska research reported here are tentative, but highly suggestive. They represent a fruitful avenue for further research in developing an approach in which knowledge concering institutional structure can be profitably applied to social impact assessment studies.

## FOOTNOTES

(1) Rank-order correlation analysis yielded an unusually high Spearman's Correlation Coefficient of .86 ($p < .001$) between 1980 population and structural differentiation. To overcome problems of multicollinearity, multiple-regression equations were computed twice for each dependent variable; once including structural differentiation and a second time including 1980 population level. Since no variables could be established, only results pertaining to differentiation and social pathology are reported here.

Population change from 1970 to 1980 did not enter as a significant variable in explaining social pathology. A potential measurement problem may have suppressed its influence. An association between rate of population increase and social pathology might have been obtained had the population data reflected a shorter period than census decades. In Alaska, a ten-year period is long enough for a community to experience several expansions and contractions of its population. This measurement problem can be alleviated in future studies now that the Alaska State Department of Labour has implemented a population model capable of identifying population trends within census decades.

## REFERENCES

Davenport, J.A. and J. Davenport, III, Eds., Boom Towns and Human Services, Laramie, Wyoming: University of Wyoming, Department of Social Work (1979).

Davenport, J. III and J.A. Davenport, (eds), The Boom Town: Problems and Promises in the Energy Vortex, Laramie, Wyoming, University of Wyoming, Department of Social Work (1980).

De Kadt, E. (ed), Tourism - Passport to Development?, New York; Oxford University Press (1979).

Durkheim, E., The Division of Labour in Society, New York, The Free Press (Original 1893) (1933).

Geisler, C.C., R. Green, D. Usner and P.C. West, Indian SIA: The Social Impact Assessment of Rapid Resource Development of Native Peoples. Ann Arbor, Michigan: University of Michigan, Natural Resources Sociology Research Lab, Monograph 3 (1982).

Guttman, L., The basis for scalogram analysis, in: S.A. Stouffer, L. Guttman, E.A. Suchman, P.F. Lazarsfeld, S.A. Star and J.A. Clausen, Measurement and Prediction. Princeton, N.J., Princeton University Press (1950).

Kim, K.D., Toward a sociological theory of development: a structural perspective. Rural Sociology, 38 (4): 462-476 (1973).

Krohn, M.D., A Durkheimian analysis of international crime rates, Social Forces, 57 (2): 654-670 (1978).

Lee, R.G., D.G. Briggs, G.F. Schreuder and J. Kronick, The local impacts of forest industries: Approach and methodology. Report prepared for the United Nations Food and Agriculture Organization. College of Forest Resources, University of Washington, Seattle, Washington (1984).

Miley, J.D. and M. Micklin, Structural change and the Durkheimian legacy: a macrosocial analysis of suicide rates. American Journal of Sociology, 78 (3): 657-670 (1972).

Muth, R.M., Identifying social effects in forestry decisionmaking: lessons from the past and prospects for the future. In: Yarie, Sally, ed., Proceedings of the Alaska Symposium on the Social, Economic and Cultural Impacts of Natural Resource Development. Fairbanks, Alaska: University of Alaska, Department of Conferences and Institutes (1983).

Muth, R.M., Structural Differentiation and Community Growth: A case study of natural resource development and social change in selected Alaskan communities. Ph.D. dissertation, University of Washington, Seattle, WA. (1985).

Nisbet, R.A., The sociology of Emile Durkheim. New York, Oxford University Press (1974).

Summers, G.F. and K. Branch, Economic Development and Community Change. Annual Review of Sociology 10: 141-166 (1984).

Webb, S.D., Crime and the division of labour: testing a Durkheimian model. American Journal of Sociology, 78 (3): 643-656 (1972).

Wilkinson, K.P., J.G. Thompson, R.R. Reynolds, Jr. and L. Ostresh Social disruption and rapid community growth: an examination of the "boomtown" hypothesis, Paper presented at the annual meeting of the Rural Sociological Society, Ithaca, New York, August, 1980.

Wilkinson, K.P., J.G. Thompson, R.R. Reynolds, Jr., and L.M. Ostresh, Local social disruption and western energy development - a critical review. Pacific Sociological Review 25 (3): 275-296 (1982).

Wilkinson, K.P., R.R. Reynolds, Jr., J.G. Thompson and L.M. Ostresh, Violent crime in the western energy-development region: Sociological Pespectives, 27 (2) 241-256 (1984).

Young, F.W., A rural development inventory. Cornell Rural Sociology Bulletin Series no. 83 Department of Rural Sociology, Cornell University (1977).

Young, F.W. and R.C. Young, Comparative Studies of Community Growth. Rural Sociological Society Monograph nr. 2. Morgantown, West Virginia: West Virginia University Bookstore (1973).

## PART D : COMMUNITY PROCESSES AND SOCIAL PARTICIPATION

D1 : Boer, J. de,
Community Response to Soil Pollution: a Model of Parallel Processes

# COMMUNITY RESPONSE TO SOIL POLLUTION
# A MODEL OF PARALLEL PROCESSES

Joop de Boer*

## ABSTRACT

A field study investigated the perception of environmental hazards in residential areas polluted by chemical wastes. Soil pollution is an important problem in The Netherlands. The government expects that in the period from 1982 to 1997 some 1000 sites will need remedial action. Of these sites about 50 concern pollution in residential areas.
In four cases of soil pollution experiences of the inhabitants, the information process, social and medical assistance and participation by the inhabitants are studied. The data were collected in comprehensive interviews with involved professionals, the relevant authorities and 756 inhabitants. The cases were in different stages of technical investigation and remedial action. Differences in managing approach, nature of the pollutants and age of the houses were related to the perceptions and expectations of the inhabitants. It will be argued that increased understanding of individual reactions to cases of soil pollution could have important consequences for policy. Furthermore, our findings suggest a number of aspects that should be included in impact assessment studies.

## KEY WORDS

Psychological and social problems, chemical wastes, communication processes, citizen participation

---

*Institute for Environmental Studies Free University.
P.O.Box 7161, 1007 MC Amsterdam, The Netherlands

This paper is one in a series of studies concerned with psychological and social problems in residential areas polluted by chemical wastes. The first aim of this project was the formulation of policy recommendations. Our findings suggest a number of proposals dealing with the apparent conflicts between authorities and residents. In order to alleviate the psychological and social effects of this environmental problem we stress the importance of communication processes and citizen participation in finding a solution.

Presently, we are engaged in extending both the empirical and theoretical base of our recommendations. A follow-up of our survey, to be carried out in the near future, should enable us to develop a model for human reactions to this type of environmental hazard. The present paper gives a brief overview of our approach by specifying a model of parallel processes dealing with the reactions of the residents to the pollution problem.

## Scope of the problem

Soil pollution is a serious problem especially in a country as densely populated as the Netherlands. A special law was passed in order to deal with former dumps of toxic waste. The government expects that in the period from 1982 to 1997 some 1000 sites will need remedial action. Of these sites about 50 concern pollution in residential areas. Decisions on environmental management may have far-reaching consequences for people living at or near these sites. Apart from having to cope with the possible health consequences, residents are confronted with uncertainty about the future of their neighbourhood. For example, the frequent mention of property devaluations lead to a situation in which residents fear to be trapped. Their only way out seems to be either the full clean-up of the site or their relocation. According to the law both options are possible but not always feasible in view of their costs.

## Some features of soil pollution

Generally, soil pollution can be regarded as environmental stressor. As compared with noise or air pollution, however, soil pollution has some distinguishing features. At least three of these should be mentioned here.

First of all it is a well-known fact that both laymen and experts have great difficulty in understanding the risks implied by soil pollution. They have to deal with uncertainties about chemical substances, concentrations, ways of exposure, uptake by receptors, and ultimate effects. Because oilfactory and visual cues are usually lacking doubts may arise as to the boundaries of the site and the number of residents which are involved.

Secondly, it has to be emphasized that soil pollution and its effects, have an episodic character. Usually there is a sudden onset of a stream of events (e.g. Levine, 1982). After the pollution has been recognized by the authorities as being in excess of specified standards, it may take at least some years to decide on remedial actions. Among other things, progress depends on technical and administrative factors. In the meantime the residents are confronted with unforeseen changes of both their physical and social environment. The attitudes of residents towards authorities and of authorities towards residents are likely to undergo changes as a result of their interaction.
This process resembles the impacts of construction projects (e.g. Finsterbusch, 1980), but the management of soil pollution involves much more uncertainty and potentially more extreme and more personal consequences.

Thirdly, it is important to note that soil pollution does not entail any actual benefits for the parties involved in the decision. The residents are confronted with new risks devoid of any benefits. Moreover they lack the freedom of choice that might induce them to accept the situation in a normal way.

Although some of these impacts might just as well be the result of a natural disaster, it is clear, however, that soil pollution is a men-made problem for which agencies or people are to blame.

## A model of parallel processes

Table 1 presents a brief overview of the various processes that are relevant to public reactions to soil pollution. In distinguishing soil pollution from other kinds of environmental stressors three features have been mentioned that point to a number of theoretical orientations. Among other things these involve theories on decision making (e.g. Janis & Mann, 1977), risk perception (e.g. Kahneman, Slovic & Tversky, 1982), defensive mobilization (e.g. Tilly, 1978), social cognition and personal control (e.g. Fiske & Taylor, 1984), and emotion (e.g. Leventhal, 1984).

We opted for a model of parallel processes for the following two reasons. As to the first, it is important to take into account the sequence of steps concerning the management of soil pollution by the authorities. The reactions of the residents are more or less tied to the main action stages planned by the authorities. The second reason concerns the necessity to distinguish appropriate levels of reactions. Although the primary focus of our research has been the elaboration of psychological processes, it should be stressed that community response involves social processes as well.

Table 1 presents some key-elements of the model. It is based on both theoretical considerations and a preliminary analysis of research data. The proposed chains of reactions should be interpreted with caution, however, as longitudinal data are not yet available. Moreover, it should be emphasized that the various stages of the model are not discrete (i.e. both residents and authorities are continuously looking forward and back during the process). The model, therefore, should be interpreted as reflecting a continuous process in which residents focus on different salient actions, cognitions and feelings at various stages.

The top row of table 1 concerns the actions planned by the authorities in accordance with the law. Without going into details these actions may be summarized as follows. Firstly, the problem has to be recognized through a global assessment of pullution levels. In excess of specified standards pollution gives cause for a closer investigation into toxicological features ans spatial dimensions of the problem. This investigation should help the authorities to decide whether remedial actions are necessary. If this is the case an investigation will be held into both the feasibility and the costs of remedial actions. After all these, time-consuming, procedures the authorities should reach a definite decision on environmental management. In view of the sanitary conditions some residents may be relocated, the site may be cleaned up, or the pollution may be insulated from the environment.

The second and the third row of table 1 concern the reactions of individual residents. In accordance with the theory of Leventhal (1984) two parallel systems are proposed, i.e. emotional and problem-oriented responses to the events. The two psychological systems are seen as independent, but in constant interaction at all stages. Some key processes will be discussed briefly in connection with the reactions at the community level mentioned in the bottom row of the table.

## Reactions of the residents

Confronted with information about the pollution of their neighbourhood the residents have to appraise the risks and the necessity of various preliminary adjustments. For example, it may be wise to curtail the consumption of food (fruit and vegetables) from private gardens and to restrict the playground of little children. At the same time the residents have to cope with both anxiety concerning possible exposure in the past and uncertainty about future developments. This stressful situatiion promotes affiliation as well as social comparison of emotional and problem-oriented responses among neighbours.

Having lost control over their living conditions the residents will sooner or later focus their attention on the causes of their troubles. Some agencies or people are to blame for the pollution. Usually the legal and moral responsibility is attributed to the authorities. On the one side they are more or less accountable for things that happened in the past, on the other hand they are responsible for any remedial action in the future. Anger and distrust are the corresponding emotions of this attribution and their intensity depends heavily on the signs of acknoledgement given by the authorities (i.e. more acknowledgement is accompanied by less anger and distrust). The organization of a neighbourhood committee reflects the collective efforts to restore control over living conditions through, among other things, the compilation of information, the regulation of contacts with the authorities, and the pursuit of pressure upon them.

When it gradually becomes clear how the authorities define the problem and which remedial actions they will select, the residents get involved in the appraisal of strategies and implications. Equity is one of the main themes in this stage, especially if the site is not expected to be cleaned up completely. Some residents have to cope with fears of being trapped and others may have reasons to hope that they will be relocated. Even the residents whose streets border on the site may get involved. In general this means that differing interests are articulated. New conflicts may arise between the residents and the authorities, but also between local and national authorities, and between different groups of residents. The decision process will be hampered and the various tensions will last longer than necessary. Unfortunately, these delays can have important consequences for future interactions between the parties that are involved.

Once the decision has been reached, the residents have to prepare themselves for the changes that are to be expected. Depending on the outcomes of the decision and their personal experiences during the previous stages, they may have to cope with feelings of helplessness, depression and fatigue. Even if they were successful in influencing the authorities, they may still be uncertain and sceptical about the consequences of the remedial actions and the ultimate effects of the pollution. Meanwhile the growing importance of personal adjustments and the decline of collective interests contribute to a process of individualization of the community level. Nevertheless, collective actions will revive whenever the events take an unforeseen turn.

The present model is yet open ended. For the time being we have no data about the reactions of the residents to the implementation of the measures. Furthermore, one can only speculate about the long-term impacts of this environmental problem. Some crucial themes may be mentioned, however. The first of these is readjustment. It is evident that both movers and nonmovers have to rearrange and reappraise their living conditions in order to restore their prior level of well-being. The second theme concerns "emotion control" in periods of vulnerability. It may turn out that some residents will find themselves being even better equipped to cope with stressful situations than before. Other residents, however, may show all kinds of malaise indicating an increased vulnerability to life-events. The third theme centers around the pattern of migration in the neighbourhood. Especially a downgrading pattern of migration will act as an additional stressor to the nonmovers.

## Some illustrative data

Our discussion of the model can be illustrated with some data provided by our field study. In four cases of soil pollution experiences of the residents,

the information process, social and medical assistance and participation by the residents are studied. The data were collected in comprehensive interviews with involved professionals, the relevant authorities and 657 residents.
The cases were in different stages of technical investigation and decisionmaking on remedial actions. Differences in managing approach, nature of the pollutants and age of the dwellings were related to the perceptions and expectations of the residents. From a social perspective, at least two types of soil pollution problems should be distinguished.

The first type concerns relatively new dwellings built on sites with deep-seated pollutants. Two of our cases fit into this type. Both cases concern home-owners in middle-class neighbourhoods. Many residents of these new dwellings have families with small children. These sites are polluted by a mixture of toxical substances that is difficult to clean up.

The second type consists of gardens polluted by chemical waste. Our two other cases fit into this type. In these cases the residents were home-owners in relatively lower-class neighbourhoods. Vegetable gardens predominate here. Most residents consumed vegetables and/or fruit from their private garden. The surface of these sites is polluted by one chemical substance. The clean-up is relatively easy, so relocation is not under discussion.

Table 2 gives an overview of the four cases in conjunction with the stages of the model. In view of the many details that should be mentioned otherwise, our illustration will be restricted to the cases of the first type. Case 1 concerns 750 dwellings built at a site that was heightened by sludge. The authorities have started a closer investigation into the pollution at the moment of our survey. Case 2 concerns 40 dwellings built at a former waste site as well as 60 adjacent dwellings. At the moment of our survey both the authorities and the residents were heavily involved in the process of decision making.

The data presented in table 3 are meant to illustrate our discussion of the model. These data refer to the main concerns of the residents elicited through a card-sorting task with a free choice out of 31 possible concerns. As to case 1 one may note that many residents are uncertain and anxious even when little is known of the pollution. Moreover they show a marked distrust of the authorities that are being accused of concealing information.

The tensions of decision making are clearly shown by the concerns of the residents in case 2. Most of their concerns center around the future living conditions in their neighbourhood. Many residents express their intention to move. While some of them have reason to hope for relocation, others fear to be trapped if the site will not be cleaned up. Although the local authorities support their case, they have serious doubts about the decisions of national authorities. The residents take an active stand in the decision making process, and they express the feeling that they have to fight even for their normal rights. Admitting their own irritability, they also mention some aggression among their neighbours.

## Concluding remarks

Soil pollution is an environmental problem with serious social consequences. This paper suggests a model of parallel processes in order to specify how the mental and physical well-being of the residents is to a large extent determined by the management of the authorities.

The authorities could improve their handling of cases of soil pollution by:

a. improving the information provided to residents,
b. improving citizen participation in these processes, and
c. to acknowledge their responsibility for future living conditions in the affected neighbourhoods.

Table 1. A model of parallel processes concerning reactions to soil pollution in residential areas.

| | | | | | |
|---|---|---|---|---|---|
| AUTHORITY ACTION: | Global assessment of pollution | Closer investigation of concentration and ways of exposure | Selection of remedial actions and estimation of costs | Decisions on environmental management | Implementation |
| INDIVIDUAL PROBLEM CONTROL | Appraisal of risk and tentative adjustments | Attribution of legal and moral responsibility | Appraisal of strategies and implications | Selection of personal adjustments | Rearrangement of living conditions |
| INDIVIDUAL EMOTION CONTROL | Coping with anxiety and uncertainty | Coping with anger and distrust | Coping with hope, fear and hostility | Coping with depression and fatigue | Coping with vulnerability |
| LOCAL COMMUNITY RESPONSE | Social comparison of risk and uncertainty | Collective efforts to get control over living conditions | Articulation of differing interests | Individualization | Possible downgrading migration |

Table 2. Some characteristics of the four cases.

| | | | | | |
|---|---|---|---|---|---|
| AUTHORITY ACTION: | Global assessment of pollution → | Closer investigation of concentration and ways of exposure → | Selection of remedial actions and estimation of costs → | Decisions on environmental management → | Implementation |
| TYPE 1 | | Case 1 | Case 2 | | |
| DEEPSEATED POLLUTANTS | | 750 new dwellings build on sludge; middle class neighbourhood with many little children | 40-100 new dwellings build on waste site; middle class neighbourhood with many children | | |
| TYPE 2 | | Case 3 | | Case 4 | |
| POLLUTION AT THE SURFACE | | 200 dwellings; vegetable gardens polluted by lead; lower class neighbourhood; middle-aged | | 100 dwellings; vegetable gardens polluted by cadmium; lower class neighourhood; old-age | |

Table 3. Some concerns of residents confronted with soil pollution (% selecting these themes in a card-sorting task).

| | CASE 1 STAGE OF CLOSER INVESTIGATION (N = 288) | CASE 2 SELECTION OF REMEDIAL ACTIONS (N = 95) |
|---|---|---|
| I worry about how they will deal with the pollution | 68% | 79% |
| I am uncertain about the health of my children | 56% | 44% |
| I don´t trust the authorities any more | 50% | 43% |
| They don´t tell us everything | 74% | 46% |
| My property is strongly devaluated (% of home-owners) | 93% | 88% |
| I will never be able to leave this dwelling (% of home-owners) | 51% | 48% |
| I should try to get another dwelling | 16% | 68% |
| I feel more nervous and | 10% | 36% |

| | | |
|---|---|---|
| irritable | | |
| I feel as if I have to fight for my normal rights | 27% | 54% |
| Some of my neighbours get aggressive | 8% | 42% |

## REFERENCES

Finsterbusch, K. (1980). Understanding social impacts. Sage, London.

Fiske, S.T., and Taylor, S.E. (1984). Social cognition. Addison, Wesley. Reading, Mass.

Janis, I.L., and Mann, L. (1977), Decision making Free Press, New York.

Leventhal, H. (1984). A perceptual motor theory of emotion, in Berkowitz, L., ed. Advances in experimental social psychology, vol. 17. Academic Press, New York.

Levine, A.G. (1982). Love Canal: Science, politics, and people. Lexington Books, Lexington, Mass.

Kahneman, D., Slovic, P. and Tversky, A., eds. (1982). Judgement under uncertainty. Cambridge University Press, New York.

Tilly, C., (1978). From mobilization to revolution. Addison-Wesley, Reading, Mass.

## PART E : TECHNOLOGY ASSESSMENT/POLICY ANALYSIS

# THE USE OF SCREENING IN THE PAWN STUDY

Warren E. Walker*

## ABSTRACT

In most policy analysis studies there are a large number of alternative policies and a large number of impacts to be considered. Time and budget constraints make it impossible to calculate all the impacts for all of the alternatives As a result, such studies often include a step (which we call screening) in which a large number of policy options(tactics) are evaluated in terms of a small number of impacts in order to screen out those tactics that are clearly not attractive. The screening method is often implicit and non-scientific.

This paper describes the systematic approach to screening that was used in the PAWN study. A two stage process was used. In the initial screening step, a tactic's cost was compared with an estimate of the maximum benefit that any tactic could conceivably produce when applied for the same purpose. Only the tactics whose costs were less than the maximum benefits survived to be evaluated in the next step.

The second screening step involved obtaining an upper bound on the expected annual benefits from the tactic under various assumptions about future demands for surface water, and comparing this upper bound with a measure of the annual cost of the tactic -- the tactic's annualized fixed cost. If the annualized fixed cost was less than the upper bound on the expected annual benefits, the tactic was considered promising, and was retained for further analysis. If not, the tactic was screened out.

The benefits considered in the screening analysis were (1) reductions in agriculture shortage losses, (2) reductions in agriculture salinity losses, and (3) reductions in shipping losses caused by low flows on the waterways. The upper bound on expected annual

benefits from a tactic was obtained by taking a weighted average of the estimated benefits in four years of varying dryness. The weights were based on the probabilities of occurrence of the various types of years.
In general, we found that, unless the demand for surface water increases significantly above current levels, there are relatively few tactics that are worth considering further. In the case of increased demand for surface water, we found a number of promising tactics. Most of them were small, inexpensive regional tactics. Almost all of the large, expensive, national tactics (some having investment costs of over $ 100 million that had been under discussion in The Netherlands for many years) were screened out.

KEYWORDS

alternative, policies, screening, screening method

---

*The Rand Corporation, 1700 Main Street, Santa Monica, California 90406

## INTRODUCTION

### The PAWN Project

Although the historical Dutch water management problem has been too much water, in recent years the Dutch have had to turn their attention to the new problems of too little water and too much pollution. Like other developed countries, the Netherlands faces water shortages, water quality problems, and environmental and distribution problems because of increased population and industrialization and higher standards of living. Competition among water users has grown. These problems are intensified in the Netherlands because of the dense population, because so much of the country lies below sea level, and because the major source of surface water is the polluted Rhine River.

The severe drought of 1976 brought these concerns sharply into focus. The drought caused extensive damage. Agricultural losses alone exceeded $2.5 billion. Low river levels caused serious shipping delays and costs because ships could not navigate the waterways with full loads. And the water shortage worsened water quality problems. These events, coupled with Dutch awareness that the modern water management problem is no longer local in scope, stimulated the Rijkswaterstaat (the Netherlands government agency responsible for water control and public works) to commission an analysis on which to base a new national water management policy. Begun in April 1977, the "Policy Analysis for the Water Management of the Netherlands" (PAWN) project was conducted jointly by The Rand Corporation, the Rijkswaterstaat, and the Delft Hydraulics Laboratory.

The PAWN project was a major undertaking. It directly involved over 125 man-years of effort, created several dozen computer programs, and gathered and structured an enormous amount of data.

The primary tasks of the PAWN project were

(1) to develop a methodology for assessing the multiple consequences of water management policies,
(2) to apply it to generate alternative water management policies for the Netherlands and to assess and compare their consequences, and
(3) to create a Dutch capability to conduct further analyses of this kind by training Dutch analysts and by documenting and transferring to the Netherlands the methodology developed at Rand.

## Organizing the Alternatives

During problem formulation, we collected suggestions from many government agencies and interest groups about possible measures for inclusion in a Dutch water management policy. The alternatives were so numerous and diverse that we decided to organize them hierarchically into three levels: tactics, strategies, and policies.

A water management tactic is a single action to affect water management, such as building a particular canal or taxing a particular use. We considered four kinds of tactics, the first two directly affected the supply of water and the other two the demand.

1. Technical: additions or modifications to the current water distribution facilities (infrastructure)
2. Managerial: changes in the rules by which a particular infrastructure, current or new, is operated
3. Pricing: imposition of a charge (tax) on water use or discharge
4. Regulation: administrative or legal restructions on water use or discharge (for example, quotas)

A strategy is a combination of tactics of the same kind; for example, the technical strategy contains all changes to current distribution facilities.

A policy is a particular combination of all four kinds of strategies. A particular policy can be thought of as an overall water management policy for the Netherlands, much like a nation's economic policy.

## The Stages of Analysis

In analyzing water management policies, we had to consider several kinds of tactics, each offering many alternatives; for example, technical tactics could modify the current water distribution system in hundreds of different ways. It was impractical to evaluate the detailed consequences of all of them. We dealt with this problem by performing the analysis in a series of stages. In the early stages, we evaluated many possible alternatives in terms of a few major impacts. In the final stages we evaluated relatively few promising alternatives in terms of the full set of potential impacts. As the stages progressed, we learned how to design increasingly promising alternatives.

Figure 1 shows our general approach, indicating the stages of analysis that take place after problem formulation.

Goeller et al. (1983) describe the complete project in detail, and Goeller and the PAWN Team (1985) give a summary of the project. In the remainder of the paper we summarize a portion of the first stage of the analysis: the screening of technical and managerial tactics. Walker and Veen (1981) provide full details of this portion of the study.

## General Approach to Screening

In large policy analysis studies the number of alternative policies is usually too large to consider each in detail. As a result, such studies often include a step in which a large number of alternatives are evaluated in terms of a small number of impacts in order to screen out the unattractive ones. The output of this screening step is a relatively small list of alternatives that appear to be "promising"-- i.e., alternatives that are sufficiently sensible and beneficial that they merit a more detailed and thorough evaluation.

In the screening stage of PAWN, the alternatives examined were called tactics. The essence of screening, then, is (1) to construct a list of tactics that might turn out to be promising, and (2) to determine which of these are promising by performing broad-brush assessments using a small number of selection criteria.

Our initial list of tactics came partly from prior studies and recommendations by Dutch experts, and partly from a needs assessment carried out by the PAWN team to identify problems with the existing water management system and locations where new tactics might be useful. The selection criteria were related to water management goals and objectives identified for the various regions of the country, and to the overall goals and objectives of the system. The criteria that we used in screening included administrative feasibility, cost-effectiveness, and dominance.

The PAWN study did not attempt to evaluate or screen out tactics on the basis of their political feasibility, even though some of the tactics considered are controversial. (A few exceptions were made where the PAWN advisory group believed that implementing the tactics was politically impossible). PAWN generally operated under the principle that the degree of public acceptance for each tactic should be assessed by Dutch policymakers and then considered, along with each tactic's costs and benefits, when they compared the different alternatives to select a preferred policy.

Screening is more an art than a science. Since a limited number of tactics are considered and a limited number of selection criteria are used, it is conceivable that a tactic worthy of serious consideration will not even be considered or will be screened out, and that a tactic carried over to the final list will later be found to be unattractive.

The results from screening rest to a considerable degree on insight, judgment, and knowledge about the water management system. A discussion of the art and science of screening is given by Walker (forthcoming).

The entire process can be compared to the steps most people follow in buying a new house. There are generally too many houses for sale in the chosen city to visit each individually. However, a few basic criteria such as purchase price, neighbourhood, and number of rooms will normally reduce the number of alternatives significantly. Then more detailed criteria, such as the heating system, layout of rooms, etc., can be considered for the remaining houses. If one follows this strategy, it is possible that one's "dream house" will be screened out in the first stage (e.g., because the price is slightly too high); but the result is usually quite satisfactory.

## THE PROCESS OF SCREENING

The screening analysis concentrated on tactics designed to alleviate two of the country's most pressing water management problems: the shortage and salinity of the nation's surface water supply. The significance of specific shortage and salinity problems and how they are manifested differs in different parts of the Netherlands. To identify and analyze tactics to alleviate these problems, we divided the Netherlands into eight regions (see Fig.2). The regions, and the dominant water problem in each, are listed below.

1. North (shortage)
2. Northeast Highlands (shortage)
3. Flevoland and Veluwe (salinity)
4. North Holland (salinity)
5. Midwest and Utrecht (salinity)
6. Large Rivers and Northern Delta (neither)
7. West Brabant and Southern Delta (salinity)
8. Southeast Highlands (shortage)

The process of screening technical and managerial tactics involved the following four steps:

1. identification of tactics;
2. estimation of the costs of the tactics;
3. estimation of the benefits of the tactics for different scenarios;
4. comparison of the costs and benefits of the tactics; unpromising tactics are screened out and promising tactics are passed on for further analysis

Each of these steps is discussed below.

## Identification of Tactics

Our first task was to assess the major national and regional water management problems in the Netherlands. These efforts culminated in a description of the problems of shortage, flooding, salinity and quality. We then compiled a list of tactics designed to resolve the major shortage and salinity problems. This task required an extensive review of proposed solutions. Several general Rijkswaterstaat reports provided initial insights. We also consulted a large number of technical reports and documents dealing with tactics that were already being considered.

Our most useful sources of suggestions were the discussions and meetings we had with Dutch water management experts. Early in the project a formal working group of such experts, called the Netherlands Support Group (NSG), was organized to provide information and guidance to the screening activities. The NSG helped us to compile the list of tactics, supplied data need to perform the analysis, and commented on our results as they were developed.

In a few cases our search for tactics to solve a specific problem did not reveal any previously proposed candidates. In these cases we designed tactics and asked the NSG to comment on their feasibility and reasonableness and to suggest improvements.

This step of the screening process produced a list of 65 waterboard plans 1) and 57 other tactics that were analyzed in succeeding steps.

## Estimation of the Costs of Tactics

In order to place costs and benefits on a comparable basis, we annualized the costs. We made separate estimates of (1) investment costs (one-time outlays to build the required facilities), (2) fixed annual operating costs (the costs to staff and maintain the facilities), and (3) variable operating costs (the energy costs to operate the facility). The annualized fixed costs (AFC) of a tactic was obtained by applying a capital recovery factor of 0.10 to the investment cost and adding the fixed annual operating cost. (A capital recovery factor of 0.10 reflects a useful life of approximately 50 years and a discount rate of 10 percent.)

We generally ignored variable operating costs, because these (mainly the cost of electricity to run the pumps) were always very small relative to the fixed costs (ten percent or less). In the few cases in which operating costs represented a large proportion of a tactic's expected annual cost, we included the operating costs in the analysis. We also generally ignored long-run costs other than the previously defined fixed costs because they were small. For example, the loss functions in our analysis consider only the short-run costs to shipping from implementation of the tactics, not the long-run costs that would result from the need to change the shipping fleet; a supplementary analysis of this assumption indicated that the shipping losses with and without the long-run costs are approximately the same.

## Estimation of the Benefits From Tactics

The benefits from a tactic vary stochastically from year to year with the rainfall and river flows (the external supply of water). For example, a dry summer yields larger benefits from a tactic that supplies irrigation than does a wet summer. Thus, to determine whether a tactic is cost-effective, we want to compare the expected benefits (the average over many years of differing dryness) with the cost. But estimating the expected benefits by the usual approach of calculating the benefits for many years and then taking the average is prohibitively expensive. (On the other hand, averaging over only a few consecutive years risks seriously underestimating the expected benefits). Therefore, we used weighted averages of tactic's benefits in four years of very different dryness to produce upper and lower bounds on its expected annual benefits, and we used these bounds in our criteria for tactic acceptance and rejection.

We identified five different external supply scenarios of very different dryness, ranging from extremely dry to extremely wet: the supplies in years 1976, 1959, 1943, 1967, 1965. We did this by statistically analyzing historical data (by 10-day period) for 1933 through 1976, ranking the years in terms of 10 different indexes of dryness. We found 1976 to be the driest year, and it is generally considered so by Dutch experts. To get a worst-case external supply scenario for our analysis (albeit an artificial one), we replaced the 1976 Rhine discharges for several ten-day periods with the minimum of the discharges for the corresponding period in the three worst years for Rhine flow (one of which was 1976). We called the resulting scenario DEX (for "extremely dry").

We assigned probabilities to the scenarios (Table 1) by estimating the shortage and salinity losses for the 44 years of external supply data. We ranked the years according to their calculated losses. Because both the shortage and salinity losses for 1976 were the worst of the 44 years, this suggests that they would be exceeded one year in 44, or in about two percent of all years. Similar calculations were made for the remaining scenarios. Because we know that the DEX scenario is somewhat drier than 1976, when we needed probabilities for DEX, we used those for 1976.

Fourth, we obtained bounds on the expected annual benefits, the area under the curve of annual benefits versus the dryness of the external supply. We did this by calculating the benefits for the selected scenarios, obtaining a point on the curve for each, and then numerically integrating to approximate the area. To obtain the upper bound for expected annual benefits, we assumed that the benefits in DEX would be obtained in all years drier than 1959; the benefits in 1959 would be obtained in all years between 1959 and 1943 in dryness; and so forth. To calculate the lower bound, we assumed that the DEX benefits would be obtained in all years drier than DEX; 1959 benefits would be obtained in any year between 1976 and 1959 in dryness, and so forth. For each bound we also assumed that there would be no benefits from tactics in any year wetter than 1976.

Table 1

PROBABILITIES OF ANNUAL LOSSES EXCEEDING THOSE OF FIVE SCENARIO YEARS

| | 1976 | 1959 | 1943 | 1967 | 1965 |
|---|---|---|---|---|---|
| Shortage losses | 0.02 | 0.07 | 0.21 | 0.63 | 1.0 |
| Salinity and shipping losses | 0.02 | 0.09 | 0.13 | 0.57 | 1.0 |

Letting B(y) represent the benefits obtained in year y from a particular tactic , the formulas for the upper and lower bounds on the expected annual benefits (EAB) for shortage losses are 2)

$$EAB < .07\beta\ (DEX) + .14\beta\ (1959) + .42\beta(1943) + .37\beta(1976)$$

$$EAB > .02\beta\ (DEX) + .05\beta\ (1959) + .14\beta(1943) + .42\ \beta(1967)$$

For salinity and shipping losses they are

$$EAB < .09\beta(DEX) + .04\beta(1959) + .44\beta(1943) + .43\beta(1967)$$

$$EAB > .02\beta(DEX) + .07\beta(1959) + .04\beta(1943) + .44\beta(1967)$$

Estimation of the benefits for each of the four years was carried out using a model called the Distribution Model, which simulates the major components of the surface water system of the Netherlands (Wegner, 1981). Given information over time on how much water enters the country (from rivers, rainfall, and groundwater flows) and how much water is extracted by the various user groups, the model calculates the water flows in the major rivers and canals, the levels of the lakes, and the concentration of salt in these waters. Using this information, it calculates the agricultural shortage and salinity losses, and losses to shipping from low-water flows and lock delays.

### Comparison of Costs and Benefits

The final step in the screening process is to compare the AFC of a tactic with the upper and lower bounds on the EAB (call these UB and LB). If AFC exceeds UB, the tactic is clearly not promising. We chose to consider a tactic promising if AFC is less than UB, recognizing that, unless AFC is less than LB, the expected annual benefits may in fact be less than the annualized fixed costs. A decisionmaker who is concerned only with the expected annual benefits of a tactic will be inclined to reject the tactic if its costs are between the upper and lower benefit bounds. However, a dicisionmaker who is risk averse (i.e., is willing to pay more than the expected annual benefits to avoid a large loss in a very dry year) might still want to accept the tactic under those conditions. Thus, the list of promising tactics included all tactics whose EAB exceed their AFC, and this included some whose EAB are somewhat less than their AFC.

## THE SCREENING ANALYSIS

### Pre-Screening

The general approach to screening technical and managerial tactics was to prepare a list of potential tactics for each region and to evaluate each one, as described above. However, before preparing these lists, a pre-screening analysis was conducted to identify those regions and scenario assumptions 3) for which the costs of implementing tactics were unlikely to be offset by the expected annual benefits in terms of reduced crop losses. These regions could then be ignored in the subsequent analysis (except if they had serious problems with respect to other impact measures).

For each region and several demand and supply scenarios, estimates were made of the agriculture shortage losses that would occur (1) with the existing water management infrastructure, and (2) if the infrastructure were expanded so that there were no shortage losses in areas with access to surface water (i.e., the demands for irrigation water were fully met).

The difference between the two estimates of shortage losses (which was called the preventable losses) would be the maximum benefits that could be expected from any tactics designed to reduce water shortages in the region. An upper bound on the preventable losses was obtained by making use of the formula for shortage losses presented above. Letting L(y) be the preventable losses in year y, an upper bound on the expected annual preventable losses (UB) is given by:

$$UB = .07L(1976) + .14L(1959) + .42L(1943) + .37L(1967).$$

Table 2 presents the maximum benefits by region for two demand scenarios -- a low-demand scenario (corresponding to the current use of water for irrigation) and a high-demand scenario (representing projected future demand). The results show that unless the demand for irrigation water increases substantially, preventable shortage losses are expected to be small, so few if any tactics are needed. Thus, the next stage of the screening analysis was limited to the high-demand scenario, except for tactics involving Region 8. In addition, since the maximum benefits for Regions 3 and 6 were very low in both scenarios, no technical tactics for reducing agriculture shortage losses in these two regions were analyzed.

## SCREENING: AN ILLUSTRATION

To illustrate how the screening analysis was performed, the analysis of one of the tactics is summarized. This tactic involves building a canal to reduce agriculture salinity losses in Region 5 (the Midwest and Utrecht). Some of the country's most valuable crops are grown in this region -- the world renowned Dutch bulbs, greenhouse vegetables, and greenhouse flowers. All are quite sensitive to salt. The principal inlet point for the region's surface water is located along the Hollandsche IJssel at Gouda (Fig. 3).

Table 2
UPPER BOUNDS ON EXPECTED ANNUAL PREVENTABLE LOSSES

(In Millions of Guilders)

| Region | Low Demand | High Demand |
|---|---|---|
| 1 | 0.0 | 7.5 |
| 2 | 0.6 | 19.9 |
| 3 | 0.0 | 0.2 |
| 4 | 0.0 | 0.7 |
| 5 | 0.8 | 4.2 |
| 6 | 0.0 | 0.3 |
| 7 | 0.0 | 0.5 |
| 8 | 3.7 | 28.1 |

The proposed solution for this problem was to extract water for the region at a point further upstream, where there was no danger of contamination from North Sea salt water. The water would be transported to Gouda via a new canal to be built through the portion of the country known as the Krimpenerwaard (Fig.3).

The benefits from the canal were measured by the reduction in agriculture salinity losses in Region 5. These benefits are partially offset by increases in shipping losses caused by reduced flows on one of the country's major rivers, which would supply water to the canal. Distribution Model runs were made with and without the Krimpenerwaard Canal using the high-demand scenario and the 1976, 1959, and 1943 external supply scenarios. (No year wetter than 1943 was run because it was clear that the canal would not be needed in such years). The differences between the salinity and shipping losses in the two sets of runs (with and without the canal) were used as the benefits to be derived from building the canal. Table 3 presents the results derived from these runs, as well as the upper and lower bounds calculated from them.

The numbers in Table 3 provide several important insights. First, North Sea salt is a serious problem only in an extremely dry year such as 1976. Second, the increased losses to shipping from this tactic are significant. In fact, in 1959 and 1943 they outweigh the reductions in salinity losses that the tactic achieves. Finally, the upper bound on the tactic's expected annual benefits is less than its annualized fixed cost, which means that it is not a promising tactic. It was therefore screened out -- i.e., not examined further in later stages of PAWN.

## SCREENING RESULTS

Screening in the PAWN project was a simple, inexpensive, yet powerful way to reduce the effort that might have been required to evaluate many potential alternatives against scores of impact measures.

The expressed purpose of screening was to select from the large number of potential tactics a reasonably small number that merited a more detailed evaluation. This objective was achieved. But screening accomplished more than just producing intermediate results to be used as inputs to the next phase of the project. It produced insights that helped guide subsequent water resources planning in the Netherlands.

First, the pre-screening analysis showed that there are relatively few tactics for reducing agriculture shortage losses that are worth considering further, unless the demand for surface water were to increase significantly above current levels. It also showed that it would not be worth the effort to generate or evaluate any alternatives for reducing shortage losses in two of the eight regions of the country.

The insights form the screening analysis were even more valuable and somewhat surprising. Prior to the PAWN study, the Dutch had been considering several large, expensive construction projects. Almost all of these projects were screened out because the costs of construction were greater than the most optimistic assessment of the potential benefits. Savings to the Dutch government from not proceeding with those projects are estimated in the hundreds of millions of dollars.

Table 3

COMPARISON OF BENEFITS AND COSTS OF
BUILDING A KRIMPENERWAARD CANAL

(millions of guilders)

| Type of Benefit | External Supply Scenario 1976 | 1959 | 1943 | 1967 | expected Ann.benef. Upper bound | Lower bound | Annual. fixed cost |
|---|---|---|---|---|---|---|---|
| Salinity | 41.5 | 0.2 | 0.0 | 0.0 | | | |
| Shipping | (6.9) | (2.4) | (0.3) | 0.0 | | | |
| Net | 34.7 | (2.2) | (0.3) | 0.0 | 2.9 | 0.5 | 5.7 |

Except for a few very inexpensive national tactics, most of the promising tactics primarily affect a single region. This suggested that the national water management infrastructure is functioning rather well, and that future water management analysis and policymaking should focus more on regional water management problems.

For the high-demand scenario, nine inexpensive regional construction projects and five national tactics were shown to be highly cost-beneficial and were passed on to the impact assessment phase of the project. Six of these fourteen tactics were also promising for the low-demand scenario. They include two aimed at reducing agriculture salinity losses, one that will help reduce agriculture shortage losses, two that will reduce low-water shipping losses, and one that is designed to reduce the chance of flooding. The combined annualized fixed cost for all six tactics is less than \$3 million. For the high-demand scenario, in addition to the fourteen tactics mentioned above, eleven others were found promising (i.e., cost-beneficial). But they were screened out because they were each dominated by one of the other fourteen (i.e., they attack the same problem, but have higher costs and/or lower benefits).

## CONCLUSIONS

This paper has demonstrated the usefulness of screening and has presented some systematic strategies for carrying it out. Screening can significantly reduce the cost and time of carrying out policy studies involving large numbers of potential policies and large numbers of impacts. In fact, it can make it feasible to carry out a study that might appear (because of the "curse of dimensionality") to be intractable.

The PAWN study illustrated the power of screening. The prescreening analysis eliminated two of the country's eight regions form detailed analysis before any tactics were evaluated by showing that no tactics could be cost effective. The screening analysis provided a broad-brush, fast and simple analysis of about 100 tactics, which eliminated 85 percent of them from further analysis. Many of these had been under serious consideration for implementation. The analysis was convincing enough that even their original sponsors did not question the results.

## NOTES

1) Waterboard plans were plans developed by local water management authorities to increase the area eligible for installation of surface water sprinklers by building local distribution facilities to supply water to a greater area.

2) The probabilities used here are not the same as

those in the table, for those are cumulative probabilities, and what we need here are probabilities for individual years, which are obtained by differencing adjacent cumulative probabilities. For example, the probability associated with year 1959 when we are calculating the lower bound is 0.05, which is the 0.07 cumulative probability for 1959 minus the 0.02 probability for DEX (1976).

3) A scenario consisted of factors that are uncertain and whose value will be determined by outside processes (political or natural) beyond the control of water management decisionmakers. The scenario variables included the amount of sprinkler equipment that would be installed by farmers and the amount of salt that would be dumped into the Rhine River outside the Netherlands.

## REFERENCES

Goeller, Bruce F., et al., Policy Analysis of Water Management for the Netherlands: Vol. I, Summary Report, The Rand Corporation, R-2500/1-NETH, March 1983.

Goeller, Bruce F., and the PAWN Team, "Planning the Netherlands´ Water Resource", Interfaces, Vol. 15, No. 1, January-February 1985, pp.3-33.

Walker, Warren E., "The Use of Screening in Policy Analysis", Management Science, forthcoming.

Walker, Warren E., and Meinaard A. Veen, Policy Analysis of Water Management for the Netherlands: Vol.II, Screening of Technical and Managerial Tactics, The Rand Corporation, N-1500/2-NETH, 1981.

Wegner, L.H., Policy Analysis of Water Management for the Netherlands: Vol. XI, Water Distribution Model, The Rand Corporation, N-1500/11-NETH, October 1981.

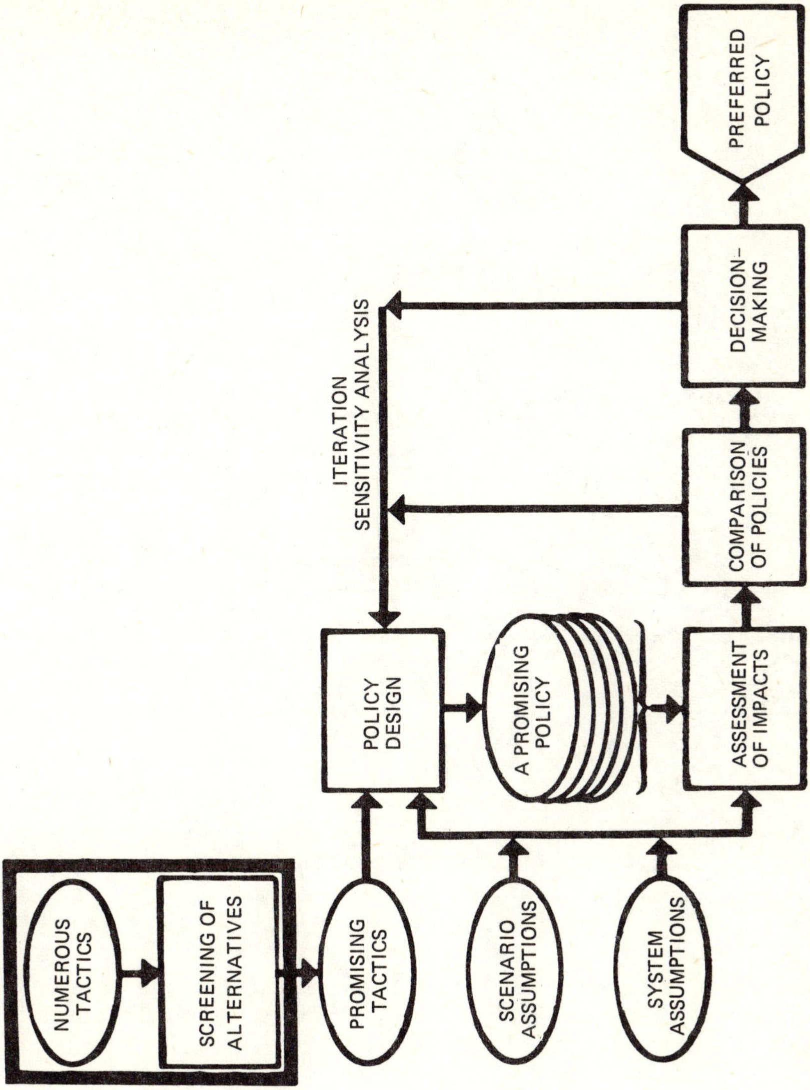


Fig. 1 -- The stages of analysis that take place after problem formulation

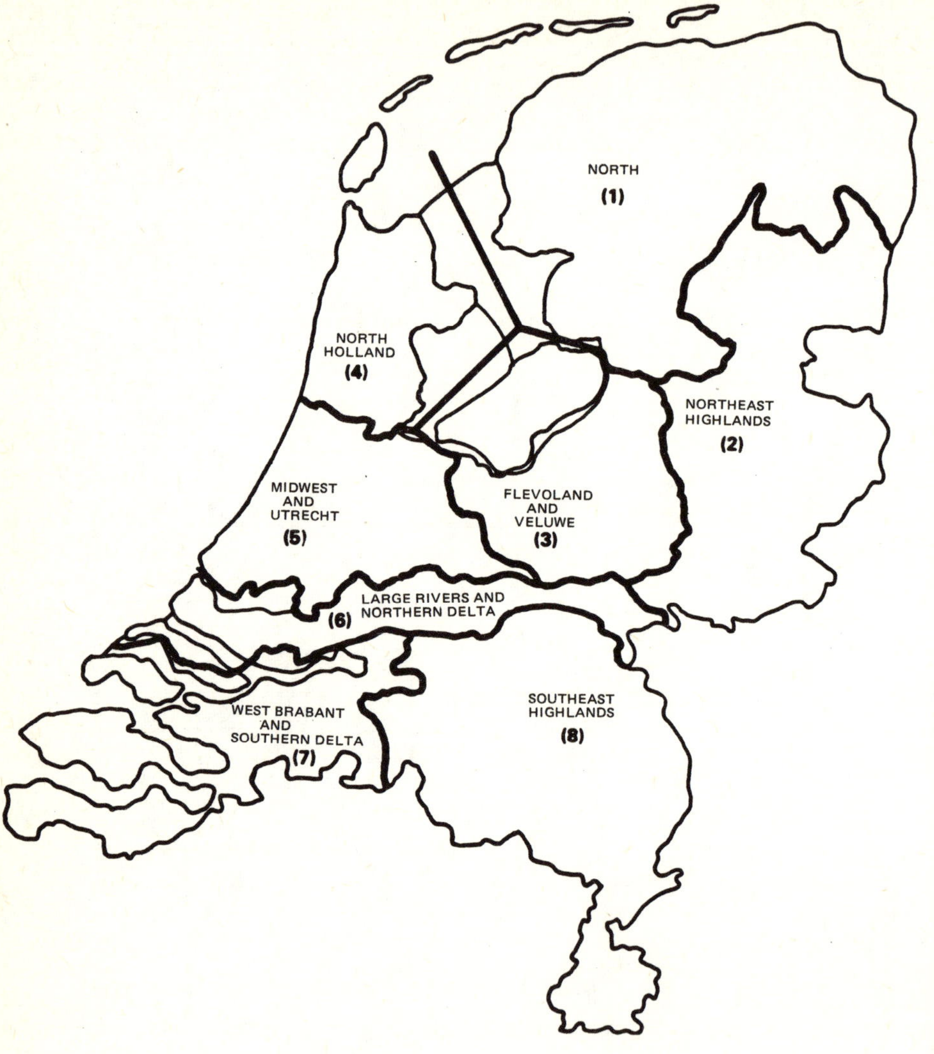


Fig. 2 -- PAWN analysis regions

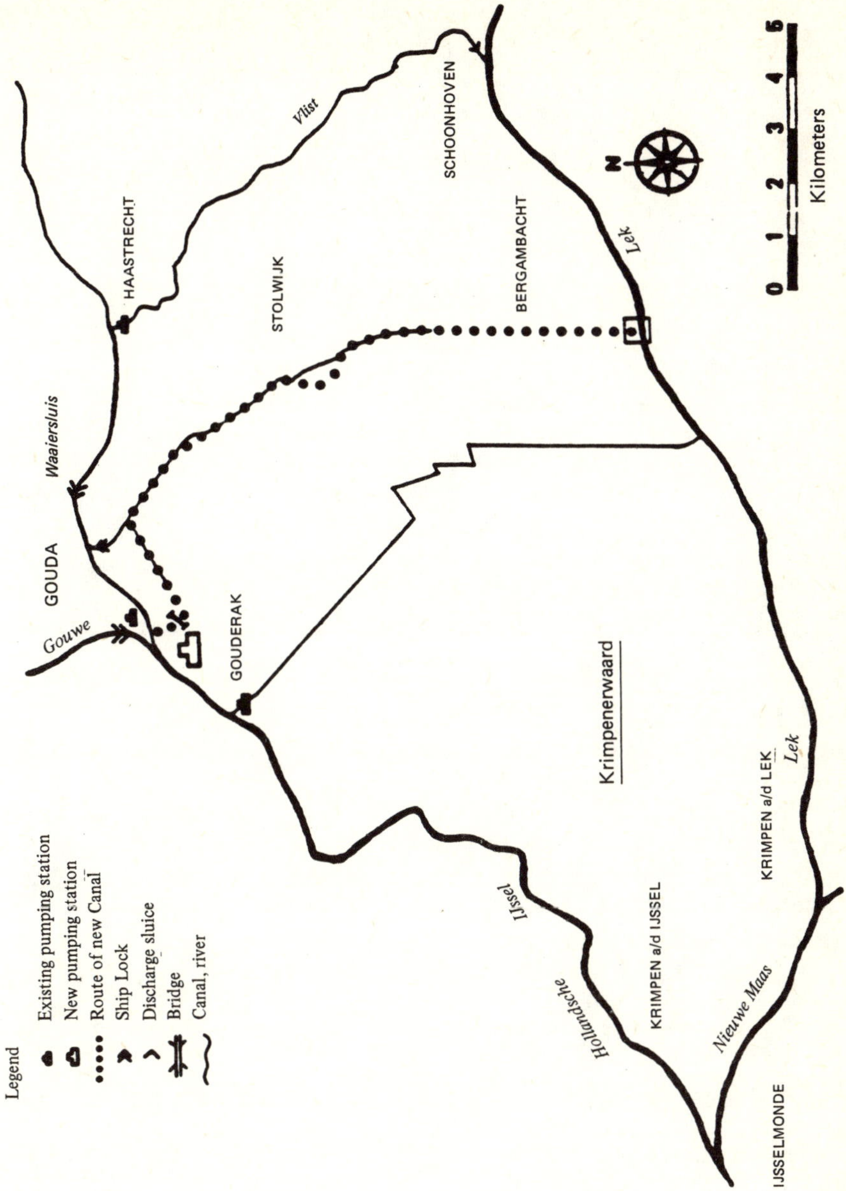


Fig. 3 -- Route of Krimpenerwaard Canal

MULTIPLE CRITERIA ANALYSIS
AND
INTEGRATED IMPACT ANALYSIS*

Peter Nijkamp**

ABSTRACT

Multiple Criteria Analysis and Integrated Impact Analysis

This paper will give a concise review of multiple criteria methods in the context of integrated impact analysis. Integrated impact assessment serves to gauge all relevant effects of external stimuli in a coherent manner. Various methods of integrated effect analysis will be reviewed with a particular view on its use in a policy context.

Next, it will be demonstrated that multiple criteria analysis may well serve the need for integrated policy analysis, based on a coherent and consistent impact assessment. A brief overview of some important classes of multiple criteria methods will be provided, based on the specific features of each method, on the nature and precision of the data available, and on the characteristics of the policy (impact) analysis at hand. Finally, an illustration of the typological approach adopted in the paper will be given on the basis of environmental policy evaluation.

---

"To press non economic values into the framework of the economic calculus, economists use the method of cost/benefit analysis. This is generally thought to be an enlightened and progressive development, as it is at least an attempt to take account of costs and benefits which might otherwise be disregarded altogether. In fact, however, it is a procedure by which the higher is reduced to the level of the lower and the priceless is given a price. It can therefore never serve to clarify the situation and lead to an enlightened decision. All it can do is lead to self-deception or the deception of others; for to undertake to measure the immeasurable is absurd and constitutes but an elaborate method of moving from preconceived notions to foregone conclusions: all one has to do to obtain the desired results is to impute suitable values to the immeasurable costs and benefits. The logical absurdity, however, is not the greatest fault of the undertaking: what is worse and destructive of civilisation, is the pretense that everything has a price or, in other words, that money is the highest of all values".

E.F. Schumacher
Small is Beautifull
A Study of Economics as if
People Mattered
Sphere Books Ltd
London 1974

---

* Various elements of this paper are based on P. Nijkamp, H. Leitner and N. Wrigley, Measuring the Unmeasurable, Martinus Nijhoff, Dordrecht, The Netherlands, 1985

** Dept. of Economics, Free University, P.O.Box 7161, 1007 MG Amsterdam

KEYWORDS

multiple criteria methods, integrated impact analysis, integrated effect analysis, policy analysis, environmental policy evaluation

## INTRODUCTION

Impact analysis is never an aim in itself, but is part of a broader policy (or decision) analysis. Consequently, impact analysis and plan (and project) evaluation are two interwoven components of policy (or decision) analysis. Plan and project evaluation has become a major element in modern public planning and administration. The history of plan and project evaluation however has not exhibited a rectilinear trajectory, but a development path marked by various phases.

The history of plan and project evaluation before World War II showed a strong tendency toward a financial tradeoff analysis. A good example can be found in the USA, where a specific emphasis was placed on the regional growth aspects of an integrated regional hydrological project.

Later on much attention was focused on the judgement of alternative American military defense systems, based _inter alia_ on cost-effectiveness principles. After World War II (social) cost-benefit analysis gained increasing popularity in public policy evaluation. The cost-benefit methodology became the leading evaluation instrument until the seventies, especially in countries with a market economy or a mixed economy. Several limitations inherent in cost-benefit analysis were relaxed by introducing amendments such as planning-balance methods and goals-achievement methods.

From the seventies onwards a new class of evaluation methods arose, called multiple criteria analysis. Many origins of the class of multiple criteria methods can be found in France. Especially the well-known concordance techniques have become a main stream in current evaluation methodology.

The following reasons can be mentioned for the increasing influence of multiple criteria evaluation techniques:

- the impossibility of including intangible and incommensurable effects in the conventional cost-benefit methodology;

- the conflict nature of modern planning problems, so that - instead of a single decision-maker - various (multi-level) formal and informal decision agencies determine a final choice;

- the shift from conventional "one-shot" decision-taking to institutional and procedural decision-making where many political aspects play a major role;

- the desire in modern public decision analysis not to be confronted with a single and "forced" solution, but with a spectrum of feasible solutions.

All these reasons have led to the current popularity of multiple criteria analysis in public planning. In the seventies and the beginning of the eighties a real avalanche of multiple criteria methods has taken place, so that at the moment there is a wide array of various multiple criteria evaluation methods. At least 40 to 50 different methods can be distinguished. This raises of course the question which methods are superior or inferior, or more precisely: which methods are suitable for which class of evaluation problems (including impact assessments).

Despite the large number of (simple and sophisticated) methods that are currently available for plan and project evaluation, there is still surprisingly little information available regarding the specific features of evaluation methods and the precise conditions under which a method is applicable.

Multiple criteria analysis aims at providing a multi-dimensional evaluation of expected impacts of a decision. Clearly, there is a wide variety of decision situations, ranging from full (administrative and financial) control over a plan or project to a marginal test or approval of a proposal. Therefore, some general elements of multiple criteria analysis will first briefly be discussed.

## ELEMENTS OF MULTIPLE CRITERIA ANALYSIS

### Introduction

Policy evaluation requires a broad judgement framework for the impacts of decisions or choices. In the past, very often a uni-dimensional measuring rod (income per capita, net social benefits) has been used, but in the recent past this uni-dimensional measure has been criticized for several reasons. This measure may at best be appropriate in a competitive system marked by full information and a fully operating price system. But even under such strict conditions many essential elements of human life (e.g., residential living conditions, quality of working life, risk, equity and national independence) cannot be translated into a common denominator whether money or not. Especially the environmental problems have made us aware of the severe limitations of the measuring rod of money.

During the last decade the insight has grown that welfare is essentially a multidimensional variable which comprises *inter alia* average income, growth, environmental quality, distributional equity, supply of public facilities, accessibility and so forth. Consequently, the welfare of countries, cities or groups should be represented by a *vector* profile instead of by a scalar. In this way, *unpriced* impacts of human activities can be taken into account as well.

Any plan or project, either in the private sphere (for example a steel plant, an office building, a shop) or in the public sphere (for example, a motorway, a local community center, a harbour) has a wide variety of impacts, not only in economic terms, but also in social and environmental terms. If public policy has to judge the social value of such projects, it should be kept in mind that ′*the* public decision-maker′ does not exist, so that due to diverging preferences and lack of concensus within a public decision agency the emergence of conflicting options cannot be avoided. No evaluation framework will solve such policy problems. Thus the meaning of evaluation for public decision-making regarding plans or projects is not primarily the identification of the optimal and unambiguous solution, but rather the rationalisation of the decision problem at hand. This means that an evaluation should focus on the provision of all relevant information on the judgement criteria (e.g., costs, employment effects, energy use, pollution effects/balance of payments), on the generation of all relevant alternatives (e.g., various locations of a shopping center), on the existence of interest conflicts (e.g., environmentalists versus supporters of economic growth), on the existence of different priorities for various impacts (e.g., employment versus budget equilibrium) and on the development of procedures and techniques which guarantee the best use of the available information, given the institutional framework (e.g., process planning, multi-level planning, interactive policy-making).

The results of an evaluation procedure may vary, and depend on the information required by the responsible public decision committee(s). Sometimes, only an assessment of all impacts (short term and long term) of a plan is sufficient, but in other cases much more emphasis has to be placed on differences in priorities or on procedural questions. Thus an evaluation framework should be flexible enough to cover all these different demands. In addition, this evaluation framework should make the choice conflicts and the consequences of choices more transparent. Finally, an evaluation problem has often a learning character, so that an evaluation framework has to be used as a tool in a process planning. This implies that - in general - an evaluation technique for alternative plans or projects should be comprehensible for the decision-maker(s) concerned, so that applicability and simplicity have to be strived for in any evaluation analysis. These conditions hold for both the provision of information on impacts or interest conflicts, and the ranking and selection of alternative investment projects.

## Information on criteria

As explained before, the social judgement of plans or projects has to be based on a broad set of criteria. These criteria may be different in nature: private economic (investment costs, rate of return, scale economies, etc.), socioeconomic (employment, income distribution, access to facilities), environmental (pollution, deterioration of natural areas, noise, etc.), energy (use of energy, technological innovation, risk, etc.), spatial-physical planning (congestion, population density, accessibility, etc.), and so forth. They may include both monetary criteria and intangible criteria, quantitative and qualitative criteria, etc. In addition to gathering the necessary data, one has to process these data in a meaningful way. Thus, an impact table may be rather detailed.

In general, human mind has a fairly limited ability to digest much information which is different in nature. Psychological experiments in actual decision situations have demonstrated that - on the average - the reliability and consistency of human decisions decreases significantly when more than seven items are to be judged simultaneously (the so-called ´magical number seven´). Therefore, it is a reasonable approach to select first a set of major important criteria (no more than seven), and next to make a subdivision into subcriteria (no more than seven), etc., so that a logical framework emerges which provides a surveyable picture of all impact criteria (see Figure 1).

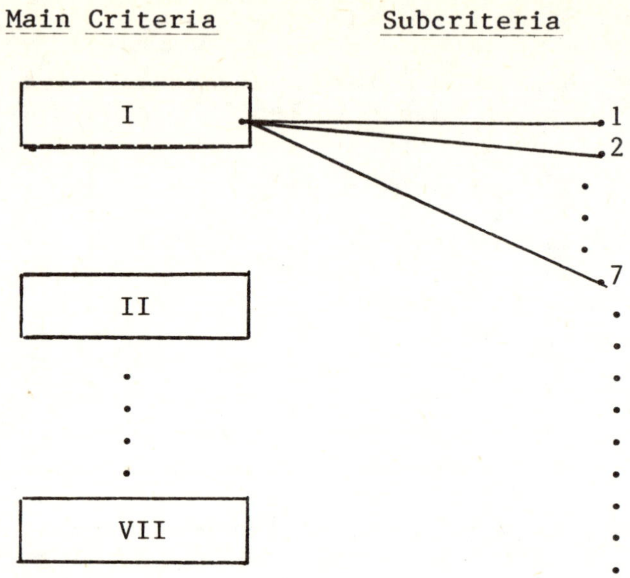


Figure 1. Subdivision of evaluation criteria into main and sub-criteria. Normally, such main criteria are: macro-economic effects, labour market effects, social effects, micro-economic effects, environmental effects, energy effects, etc. It should be added that the division into 7 main criteria and 7 sub-criteria does not mean that always precisely 7 criteria have to be distinguished; of course, the number of relevant criteria may be lower. The ultimate selection of relevant judgement criteria is a political responsibility, supported by necessary information from the side of experts or analysts. It should be noted, that the evaluation of projects in the framework of a so-called process planning often gives rise to the introduction of new (or adjusted) criteria which were neglected during the initial stages of the evaluations procedure. Therefore, the procedure should be flexible enough to include new evaluation criteria during a later stage.

The selection of judgement criteria should not primarily be based on the availability of reliable information, but rather on the relevance of these criteria in the policy evaluation framework. It will be indicated later, that also soft and qualitative information can be taken into account, so that there is no need to exclude *a priori* fuzzy information.

It should also be mentioned that the information on the criteria may either refer to a static (or comparative static) framework, in which the time dimension does not play an important role, or to a dynamic framework in which long-term impacts and sequential multi-temporal impacts play a dominant role. It is evident that the use of dynamic evaluation models requires much more information; sometimes it may be meaningful to employ a discount rate to translate all future impacts into present values.

Identification of alternatives

The evaluation of plans or projects requires a judgement of alternatives. The number of alternatives may vary between 1 (should a certain project be undertaken or not?), any discrete number (for example, 10 alternative routes of a highway) and infinity (for example, the quantity of oil to be imported). It should be noted that this subdivision of evaluation projects rests only on a technical criterion concerning the number of alternatives.

The problem of one alternative is essentially a 0-1 choice system in which a choice has to be made between the *status quo* and a new situation. Sometimes an intermediate stage has to be inserted, especially when the information is not sufficient to make a decision. In the latter case, one faces, in fact, a situation with three alternative choices: ´yes´, ´no´, ´delay in order to get more information´.

The *discrete* evaluation problem (distinct number of alternatives) is very common in normal plan and project evaluation problems, in which a choice out of a finite number of alternatives has to be made. Both cost-benefit analysis and multiple criteria analysis are addressing themselves to these kinds of discrete judgement analyses.

It is clear that the number of alternatives may

sometimes be very high. In such cases it is useful to identify first a subset of alternatives which are clearly discriminating. After a first selection of the most relevant project, one may next choose the most desirable project from the set of reasonably relevant projects.

The continuous evaluation problem is concerned with a high number of alternatives. An example is the number of barrels of oil to be imported in a certain year. This import may depend on the production and consumption in the country concerned, the technological efficiency, the search for substitutes, climatological conditions, etc. The ultimate oil import policy may be based on several criteria: costs of oil, international political uncertainities, desired degree of self-sufficiency, etc.

It should also be mentioned that the total set of alternatives to be judged in an evaluation process is essentially also a matter of political responsibility. On the other hand, the expert or analyst should also take into account certain rules for defining the alternative plans on projects to be judged:

- the alternatives should be feasible from a technical or economic viewpoint.
- the alternatives should satisfy at least certain a priori specified social minima (for instance, employment levels, pollution levels, risk, energy use).
- the set of alternatives should be in agreement with the various desires and options of groups in society and of the responsible decision committees.
- the alternatives should also be distinguished according to their degree of reversibility (for example, a superstore is easier to replace than a nuclear power plant).
- the various alternatives should be defined such that they are of a comparable size (for example, it is less useful to evaluate a new big steel plant against a small bookstore).
- the level of information (e.g., the impacts of

the criteria) should be comparable for all alternatives (for example, it is less meaningful to analyze in detail all employment impacts of investment project 1, whereas the impacts of project 2 are only studied in a loose way).

The ultimate combination of criteria and alternatives gives rise to the following impact matrix, which has to be measured in appropriate units:

| criteria \ alternatives | 1 . . . . . . . . . . . . . . I |
|---|---|
| 1 | |
| . | |
| . | |
| . | |
| . | |
| . | |
| . | |
| . | |
| . | |
| . | |
| . | |
| N | |

This impact matrix may contain cardinal (metric) information (e.g., number of jobs), ordinal rankings (1, 2, 3, ....) or qualitative information (e.g., good, better, best).

The impact matrix can - without any further information about priorities - be used in several ways. In the first place, one may carry out a dominance analysis by examining whether a certain project (or a set of projects) is better or worse than the remaining ones. In this way the set of relevant projects to be evaluated can be truncated.

Secondly, one may carry out a strength - weakness analysis by identifying for each project n(n=1, high or a low contribution to the policy criteria at hand:

| Criteria | |
|---|---|
| Favourable | Unfavourable |
| 1 | . |
| 2 | . |
| 3 | . |
| . | . |
| . | . |
| . | . |
| . | I - 1 |
| . | I |

In this way one may easily identify which projects are in agreement with a certain policy criterion.

Thirdly, one may also rank all projects for each separate criterion according to the degree they contribute to that criterion. For example:

| Criterion | Rank order of investment projects |
|---|---|
| 1 | Plan 1 > plan 3 > plan 7 ........ |
| . | |
| . | |
| . | |
| . | |
| . | |
| . | |

## Identification of conflicts

The choice of plan or project out of a set of alternatives is often characterized by conflicting options. Such conflicts may be of an economic or technical nature (e.g., a conflict between the balance of payment and a social security system) or may be the result of human reactions. In general, every decision committee wants to implement a plan or project which has a maximum positive impact on income and employment, whereas at the same time it does not do harm to environmental quality and to the stock of energy resources. Such an utopian situation, however, is not likely to exist, so that in general a decision-maker has to make a certain tradeoff among conflicting criteria. The analyst may help to explain the nature of such conflicts. A useful way is to construct for each alternative project a so-called conflict matrix 1, .... I, which represents the pairwise mutual conflicts between successive criteria based on a qualitative rank order:

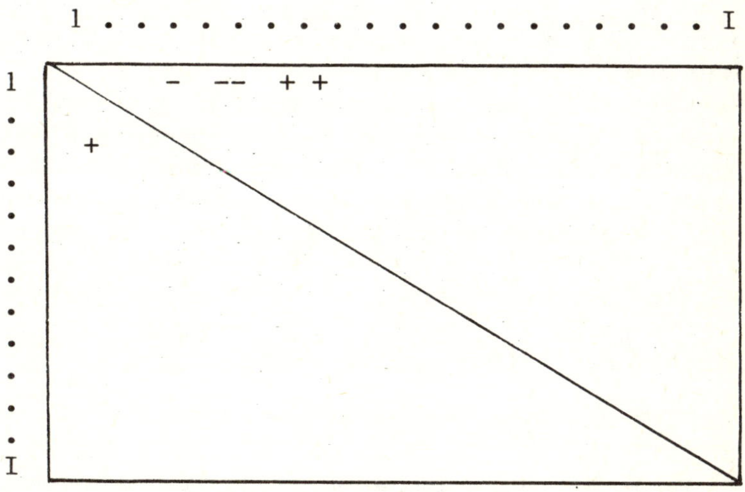


Each element of this table indicates whether or not a certain criterion is in conflict with another criterion. The symbols have the following meanings: +++ (strongly complementary), ++ (average complementary), + (weakly complementary), 0 (neutral), - (weakly conflicting), -- (average conflicting), --- (strongly conflicting).

It is often useful to disaggregate the conflict matrix for certain groups which are affected by the implementation of a certain investment project, because the welfare impacts of a given plan or project may have different effects on different socio-economic groups. Analogously, one may also carry out a spatial disaggregation in order to cover the problem of different plans or project impacts through space (for example, the impacts of a reconstruction of the inner city upon all urban districts).

One should also be aware of the fact that conflicting interests among groups may lead to a criticism regarding the set of alternatives or the set of criteria to be considered: every group wants to have reflected its interest also in the choice of the set of alternatives or the decision criteria to be judged. In such cases, the expert or analyst better takes the combinations of alternatives and criteria which are judged to be relevant by all successive groups. This choice, of course, depends also on institutional structures in decision making.

Finally, conflicts may also emerge due to lack of coordination in the decision procedure itself. Especially in the case of a multilevel policy structure or of a multiple committee structure, frictions and conflicts are very likely to arise.

The expert's task is not to solve all policy conflicts, but by means of a systematic presentation of all impacts and all bottlenecks he may contribute to a rationalisation of the decision process.

## Estimation of priorities

In as far as non-monetary values are included in the evaluation method (e.g. intantigles), any choice among alternatives is (implicitly or explicitly) based on priorities regarding the successive policy criteria for judging the plan or the project concerned. It should be noted that when priorities are expressed as weights, such weights are to be related to the units of measurement of the effects.

If the decision committee does not want to specify in advance certain priorities, the analyst's task is rather limited. In that case, (1) he has to construct in a surveyable manner the whole impact matrix, (2) he may identify all kinds of conflict inherent in the evaluation problem at hand and (3) he may examine whether certain alternatives in the impact matrix are dominant (so that these plans or projects may be subject to further analysis) or whether certain alternatives are dominated (so that they may be excluded from a further evaluation). Especially the latter dominance analysis is often extremely important, as it allows the researcher to delimit the set of plans or projects to be judged and even sometimes to rank the successive projects (or subsets therof).

If the choice among the alternatives has to be based on political priorities (weights) regarding the decision criteria, several methods can be employed to assess political priorities:

- derivation of priorities on the basis of an *ex post* analysis of decisions taken in the past. Clearly, this approach is not useful for unique decision problems.
- derivation of priorities on the basis of official documents and statements from the side of the responsible decision committee. This approach is sometimes very useful to gather information regarding general issues and policy objectives, but normally it is less useful to assess precisely preferences for refined policy criteria.
- direct assessment of priorities on the basis of interviews or questionnaires among the decision makers This is very often a fruitful method, although it is a problem that in practice many policy makers do not like to express their preferences directly.
- "fictitious" assessment of priorities on the basis of

consistent policy scenarios reflecting hypothetical but otherwise reasonable priorities for the policy criteria at hand. This is often a very useful and practical method, because it does not commit directly the policy maker, whereas on the other hand the consequences of such fictitious policy priorities can easily be traced. Such scenarios may also be derived from official documents.

It is evident that in many choice situations the information about policy preferences is fairly weak. In such cases it is always necessary to carry out a sensitivity analysis with respect to the values of political weights. Alternatively, in case of soft information one may also employ (multidimensional) scaling methods. Multidimensional scaling methods are techniques which translate soft (ordinal or qualitative) information in a consistent way into cardinal (metric) information. In that case, standard numerical operations can again be applied. Clearly, if the impact matrix contains also inaccurate or unreliable data, a sensitivity analysis should here also be applied.

Finally, it should be mentioned that different decision groups (for instance, in a multi-layer-structure) may have different priorities, so that the definition of a unique set of priorities is not always possible. In that case, it is more appropriate to assess several priority sets for the policy criteria and to calculate successively the consequences of each separate set. The results can then be further analyzed, for example, regarding the possibility to find compromise choices (see Figure).

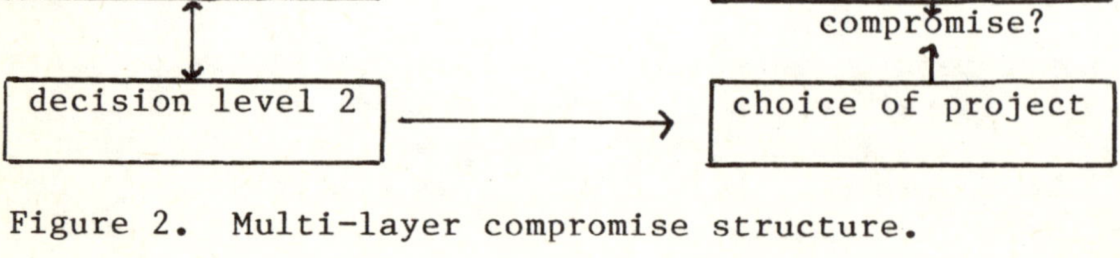


Figure 2. Multi-layer compromise structure.

## A classification of evaluation problems

Any evaluation method for judging the desirability of plans or projects should be logically and consistently connected with the nature of the decision problem concerned. Consequently, there is no unambiguous method with a universal validity. Each type of decision problem requires its own specific evaluation method.

Depending on the problems at hand and on the precision of the data used, several subdivisions can be made:

- discrete versus continuous evaluation problems
- soft versus hard evaluation problems
- static versus dynamic evaluation problems
- multi-person (or multi-committee) versus single-person (or single-committee) evaluation problems
- evaluation problems based on the generation of efficient alternative solutions versus those based on the selection of one ultimate alternative
- single-step versus process evaluation problems.

## Interactive evaluation procedures

Modern procedural or process planning requires the use of interactive evaluation procedures. Interactive evaluation procedures are based on an information exchange between the analyst and the decision committee, especially for situations in which the decision committee has not specified its preferences or weights. The basic feature of such procedures is that the analyst suggests a certain provisional feasible compromise solution to the decision committee, while next this committee may respond to this suggestion by indicating which policy objectives in this provisional solution are unsatisfactorily fulfilled. This provisional solution may be calculated by means of a standard compromise algorithm.

Thus, instead of presenting directly a final

solution, the analyst has to develop a learning procedure in order to reach in a limited number of steps a satisfactory final compromise solution. Consequently, the first compromise solution is only a trial solution which has to be presented to the decision maker(s) as a frame of reference for judging alternative efficient solutions. The easiest way to carry out such an interactive procedure is to ask the decision maker(s) which values of the policy objectives are satisfactory and which are unsatisfactory (and hence have to be improved).

This can easily be done by using a checklist which includes all values of the first compromise solutions of the policy objectives:

| Values of provisional compromise solutions | satisfactory Yes | satisfactory No |
|---|---|---|
| Policy criterion 1 | | |
| Policy criterion 2 | | |
| . | | |
| . | | |
| . | | |
| . | | |
| . | | |
| Policy criterion I | | |

Suppose now that there are K criteria ($K < I$) which are judged as unsatisfactory by the decision maker(s). This implies that a new solution has to be identified which is such that these criteria lead to a better performance. Consequently, all solutions which have a worse performance may be eliminated. After such a truncation of the set of feasible solutions a new provisional compromise solution may be calculated which can be dealt with in the same way, until finally a convergent satisfactory solution has been reached. The steps of such an interactive procedure are presented in Figure 3.

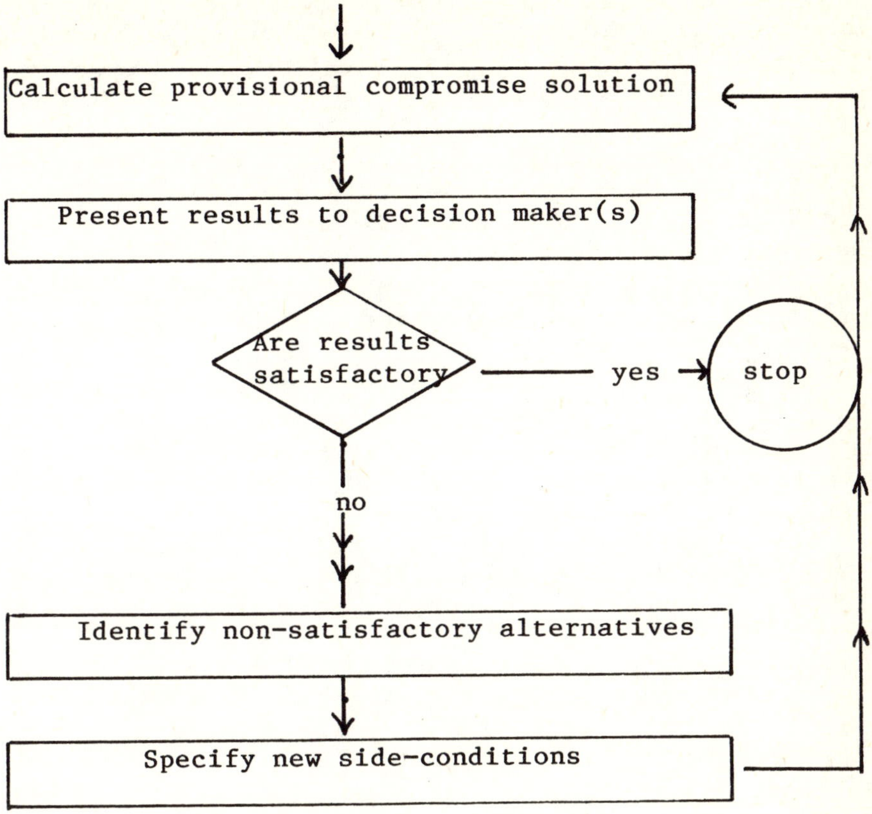


Figure 3. Interactive evaluation procedure

The benefits of such interactive procedures are: they provide information to the decision-maker(s) in a stepwise way, they can easily be included in a processplanning, they lead to an active role of the decision-maker(s) and they avoid the prior specification of ´numéraires´, preferences or weights (though they can be inferred *ex post*).

Conclusion

The appealing characteristic of the abovementioned framework for evaluating public projects is the way in which diverging impacts and conflicts are integrated. Instead of a narrow monetary view of impacts, these evaluation methods provide much broader information on, and insights into the nature of real-world choice problems. It should be mentioned, however, that these new methods are not in contrast with traditional evaluation techniques such as cost-benefit and cost-effectiveness analysis. They can easily be integrated with a cost-benefit frame-work; they are to a certain extent a natural follow-up of cost-benefit analyses, especially because latent conflicts and intangible effects can be taken into account. Such modern evaluation methods are more acceptable for the practice of decision-making by integrating decision-makers and experts right from the beginning in the evaluation process. This may also lead to a more structured decision process, since the inventory, classification, valuation and judgement of different project impacts is built up in a systematic way, based on well defined policy criteria and presented in a comprehensible way.

## A REVIEW OF SOME MULTIPLE CRITERIA METHODS

### Introduction

As set out before, the aim of multiple criteria methods is to provide a rational basis for classifying a number of choice possibilities (for instance, alternative policies, plans, neighbourhoods, regions, etc.) on the basis of multiple criteria. These methods often use two kinds of input data, viz. an impact matrix and a set of political weights attached to the criterion effects.

The impact matrix (sometimes also called evaluation matrix or project-effect matrix) will be denoted by the symbol P. This matrix has elements $p_{ij}$, which represent the impact of alternative i (i=1, ..., I) on the value of criterion j (j=1, ..., J). In the case of a qualitative evaluation problem, $p_{ij}$ may be measured on an ordinal, binary or nominal scale. However, it is not unusual that a part of $p_{ij}$ elements are quantitative in nature, i.e., some of the criterion effects are determined on a cardinal scale, whereas other effects are represented in a qualitative way. This is called a mixed data problem.

The set of weights provides information on the relative importance attached to the outcomes of the succesive J criteria; they will be denoted as a vector $\underline{\lambda}$:

$$\underline{\lambda} = (\lambda_1, \ldots, \lambda_J)' \qquad (1)$$

Clearly, usually the vector does not contain cardinal tradeoffs, but ordinal or binary weights. Most recently developed multiple criteria methods take explicit account of the "soft" nature of weights. This will be discussed in more detail in the next sub-sections.

Dominance and frequency analysis

Dominance and frequency analysis are both able to deal with a qualitative impact matrix. A dominance analysis has a very simple structure. This approach concerns the identification of alternatives that provide - in regard to all criteria - better outcomes than remaining alternatives. Consequently, only the dominating alternatives have to be considered further while all others may be eliminated.

A frequency analysis is somewhat related to this approach. The outcomes of the impact matrix are subdivided into 3 categories, viz., high outcomes (H), intermediate outcomes (M) and low outcomes (L). In a similar way, the weights may be classified into: important (T) or unimportant (O). By systematically classifying all alternatives into the successive resulting combined impacts - weights classes, a strength-weakness analysis may provide insight into the ´strong´ and ´weak´ alternatives. This strength-weakness analysis can be extended towards a probabilistic-oriented approach by calculating the frequency of scores of the alternatives in the combined impact-weights classes.

An alternative frequency analysis is the numerical interpretation method, which relaxes the assumption of the previous approach concerning the fixed number of impact categories. A limitation of this approach is, however, that varying impacts-weights classes have to be defined and numerically assessed, an operation which is, to some extent, arbitrary.

Ordinal concordance analysis

Concordance analysis has become a popular multiple criteria method, based on a pairwise comparison of alternatives. The central concept in a concordance analysis is the so-called concordance index $c_{ii'}$. This index represents the extent to which alternative i is better than alternative i′. This index may be defined as the sum of weights attached to the criteria included in the so-called concordance set $C_{ii'}$; this is the set of all evaluation criteria for which alternative i in the impact matrix P is at least equally attractive as alternative i′. Clearly, this set can be determined irrespective of the degree of information on the impact matrix.
Hence, the concordance matrix can be defined as follows:

$$c_{ii'} = \sum_{j \in C_{ii'}} \lambda_j$$

A dominating alternative can now be found by employing threshold values, graph theory or relative dominance indicators.

In an analogous way, one may define a discordance index. This index reflects the extent to which alternative i is worse than i′. Instead of using weights in this index (cf. (2)), the corresponding relative pairwise differences from the impact matrix are then taken into consideration. By combining the results from the concordance and the discordance approach, final inferences on the ranking of alternatives may be made.

In the case of ordinal information the calculation of the concordance index and the discordance index is more difficult, as numerical operations like (2) are not permitted. Sometimes the use of graph theory may be helpful, while also a transformation of qualitative information to size and importance classes may be made. In the latter case, the concordance and discordance indices may be based on the frequencies that - for each importance class - alternatives score with respect to a certain size or class impacts.

Despite the wide range of applications of the concordance analysis, also this method has some limitations. For instance, in the case of qualitative information this approach cannot be applied in a straightforward way due to the way weights and impact scores are used.

## Permutation analysis

The permutation analysis addresses especially the question: which rank order of alternatives is (after a series of permutations) in harmony with the ordinal information contained in P and $\lambda$ ?
In case of I alternatives, the total number of possible permutations will be equal to I! Each permutation can be numbered as p (p=1, ...., I!). In the permutation analysis, each rank order from the permutations is confronted with the ordinal information contained in each of the J columns of the impact matrix. Then Kendall's rank correlation coefficient is used in order to compute the statistical correlation between the I! rank orders and the J columns of P. This leads then to I! x J rank correlation coefficients denoted by $\tau_j^{po}$. Clearly, a certain permutation p (i.e., rank order of alternatives) is more attractive as the value of $\tau_j^p$ is higher.

The next step is to relate this information to the weighted vector $\lambda$. If $\lambda$ were cardinal, an ordinary weighted summation method might be applied in order to calculate the value of $\lambda_j^p$. In the case of ordinal information on $\lambda$ , the following programming model is used:

$$\max_{A}! \sum_{j=1}^{J} \lambda_j \tau_j^p \text{ , subject to}$$

$$\lambda_1 \geq \lambda_2 \geq \ldots\ldots \geq \lambda_J, \quad (3)$$

$$\sum_{j=1}^{J} \lambda_j = 1.$$

The constraints in (3) reflect the available information about the $\lambda_j$ and are used as follows. First, all extreme points that are in agreement with the constraints on the weights are generated: (1, 0, ,,,,, 0), (1/2, 1/2, 0, ...., 0), (1/3, 1/3, 1/3, 0, Next, a set of combinations of weights reflecting interior points of the above-mentioned extreme area are generated. Then one may calculate for each of these combinations which permutation p yields the maximum value of the objective function in (3). On the basis of this information, one may try to identify a rank order that is in agreement with the ordinal information from P and $\underline{\lambda}$.

This approach also has some limitations. For instance, the weights $\underline{\lambda}$ are dealt with in a rather unusual way by relating them to the coefficients $\lambda_j^p$ instead of to the criterion impacts. This implies that the size of the weights cannot be linked with the size of the impact scores, which may make them almost meaningless to the user(s).

## Multidimensional scaling analysis

A scaling technique aims at transforming a qualitative data input (mainly ordinal data) into a cardinal input of lower dimensionality. In a sense, a scaling technique may be regarded as a qualitative principal component analysis. The major principles are mainly developed in the field of psychometrics. In the recent past, scaling methods have found many applications in the area of planning, regional economics and geography.

It is clear that several concepts from scaling analysis may also be applicable to ordinal multiple criteria problems. Various approaches can be imagined in this case. In the first place, one may use a scaling technique in order to transform a qualitative impact matrix into a cardinal matrix with less dimensions. Then the cardinal configuration of the initial qualitative matrix provides a metric picture of the Euclidean distances both between the alternatives and between the effects. This is a normal standard operation.

Secondly, one may also apply a scaling analysis jointly to a qualitative impact matrix and a qualitative weight vector. In that case, both the impacts and the weights have to be transformed into a cardinal metric scale. Though this is mathematically fairly difficult, one may ultimately arrive at cardinal results for both impacts and weights. The final result of this analysis is that one is able to indicate precisely which rank order of alternatives is consistent with a certain rank order of ordinal weights.

In advantage of this kind of multiple criteria analysis that mixed data input can be treated, i.e., impact matrices containing both qualitative scores and qualitative rankings. However, a limitation of this elegant but complex approach is that it requires a sufficient number of degrees of freedom to allow a multidimensional scaling. This implies that unless sufficient impact effects are specified, no scaling results can be obtained.

## Regime analysis

Regime analysis bears a certain resemblance to concordance analysis as described in section 3.3. The starting point of regime analysis is the concordance index $c_{ii}$ defined in (2). Given the definition of this index, $c_{ii'} - c_{i'i}$ can be interpreted as an indicator of the relative attractiveness of alternative i compared with i′. Since it is assumed that the weights $\lambda_j$ are ordinal, it is impossible to find a unique numerical value for $c_{ii'} - c_{i'i}$. Therefore regime analysis focusses on the <u>sign</u> of this indicator rather than on the size of it. It can be shown that in certain cases, ordinal information on weights is sufficient to determine this sign, so that a final ranking of alternatives can be derived from the pairwise comparison matrix, consisting of values +1 and -1. In other cases, this sign cannot be determined unambiguously, however. It can be shown that in such cases a partitioning of the set of cardinal weights can be derived, being in agreement with the ordinal information on weights, such that for each subset of weights again the sign of $c_{ii'} - c_{i'i}$ can be determined.

The final result of the method is a complete and transitive ranking of all alternatives, for each of the above mentioned subsets of weights. In addition, the method produces the relative size of each subset so that one knows the relative importance of each subset.

## Evamix approach

In order to cope with the limitations of multidimensional scaling, and in order to be able to deal with mixed information, another set of methods has been developed to deal especially with <u>mixed</u> qualitative-quantitative impact scores. These methods are all based on one and the same procedural framework, which is named the <u>Evamix approach</u>.

This approach involves the construction of two measures: one only dealing with the ordinal criteria and the other one dealing with the qualitative criteria. By making various assumptions about standardization and aggregation, an appraisal score for each alternative can be calculated. The global structure of the Evamix procedure is summarized in Figure 4.

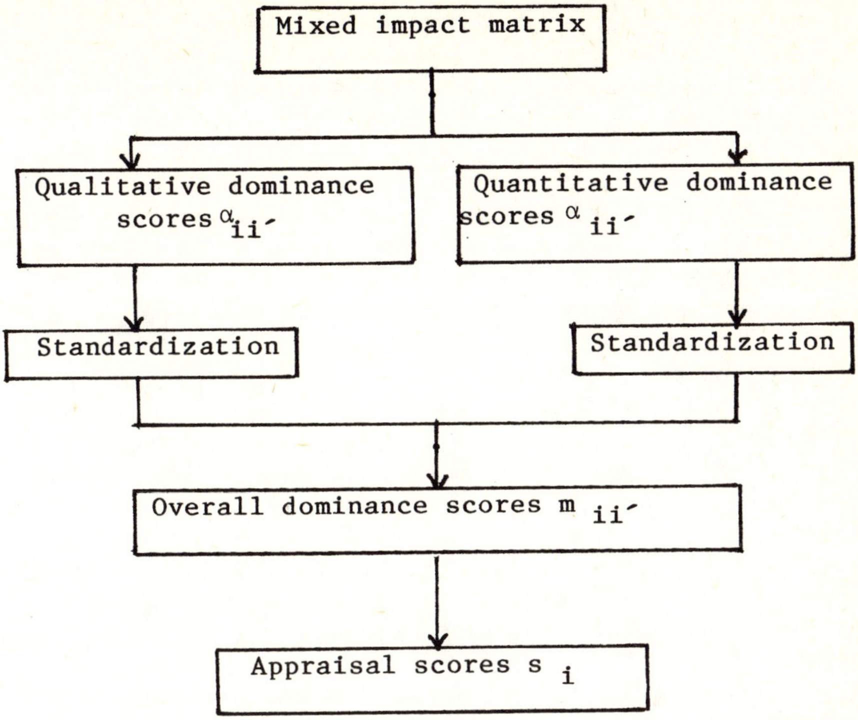


Figure 4. A brief overview of the Evamix procedure.

Thus different methods have been constructed using different sets of assumptions. Consequently, by simultaneously applying these methods to an evaluation problem, insight can be gained into the sensitivity of the final results with respect to "technical" matters. This is extremely important, especially if the evaluation results are directly used by non-experts.

## Qualitative multiple objective choice models

In contrast with the wide variety of cardinal multiple objective choice models, there is only a limited number of qualitative continuous multi-dimensional choice models. The following approaches may be distinguished.

- Fuzzy set models
- Stochastic models
- Soft econometric models

In general, <u>fuzzy set methods</u> deal with criteria of constraints that are not sharply defined, so that the boundaries of the decision area are not marked out in an unambiguous manner. By means of fuzzy set theory one may derive measures for the extent to which an element (aspect, e.g.) belongs to a certain class. This qualitative measure is usually called a membership function. Clearly, the definition of membership functions is essential in using fuzzy set approaches.

<u>Stochastic approaches</u> in continuous multiple objective models have been rather uncommon up to now. Nonetheless, if qualitative or ordinal variables can be approximated by cardinal (metric) variables with a certain probability distribution, one may construct a stochastic optimization model.

Finally, the <u>soft econometric approach</u> involves a transformation of qualitative or ordinal data input into metric units (by using the above-mentioned multidimensional scaling methods) which may then be analyzed further by means of an appropriate cardinal multiple objective method.

## Selection of an appropriate evaluation method

In this section, the following problems will be addressed: which evaluation method from a series of potential methods is applicable in a specific context? This will be illustrated by means of an example from environmental management.

Environmental management problems are glaring examples of unpriced and conflictuous decision problems which may be analysed by means of multiple criteria decision techniques. However, not all multiple criteria methods are suitable for all environmen-

tal management problems. Therefore, to research lines have to be followed:
(i) The identification of a set of activities (traffic, resource extraction, industry, etc.) which are connected with environmental problems and/or policies), and (ii) the identification of a requirements profile for the evaluation method in regard to each class of activities.

By combining the activity vector with the requirements profile, one may examine whether or not for a certain class of activities one or more appropriate evaluation methods are available.

Evaluation problems can be distinguished inter alia on the basis of the following features (see figure 5).

- the attributes of the activities to be evaluated (for instance, the construction of a motorway or the introduction of a so-called "bubble"-policy)
- the characteristics of the effects caused by the activities (for instance, local or spill-over effects, short-term or long-term effects)
- the nature of the decision structure related to the activity (for instance, a hierarchical institutionalized policy structure, participatory decisions, etc.).

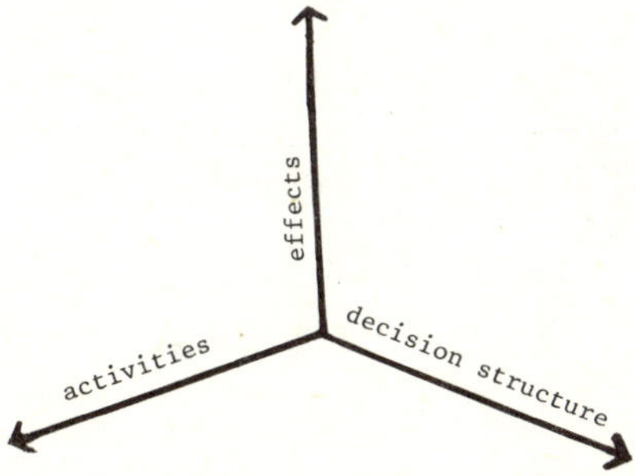


Figure 5. Three dimensions of environmental evaluation problems.

In the framework of this illustration of environmental management issues, activities will only be included if:
-these activities cause a relevant environmental

impact
-the government may exert an influence on the implementation of these activities (changes, regulations, etc.).

Furthermore, beside environmental effects, the non-environmental consequences will only be included if they play a role in the trade-off of the environmental policies concerned.

The activities profile

The identification of classes of activities takes place on the basis of comparable environmental effects and comparable policy measures.

The following <u>classes of activities</u> have been distinguished:

1. transport and transport infrastructure
2. land use and reconstruction of rural areas
3. urban land use, building and reconstruction
4. urban management and use
5. resource extraction
6. waste disposal
7. industry
8. environmental upgrading.

A further subdivision of these activities can be made on the basis of the <u>nature</u> of these activities and of their <u>spatial</u> scale.

The <u>nature</u> of these activities is determined by the range of effects, the level of precision and the extent of policy intervention. Three categories may be mentioned here:

- <u>Projects</u> The activity can be clearly identified and demarcated in space and time (for instance, the construction of an industrial plant). Projects may be further subdivided into:
  - direct governmental influence (for instance, railway construction)

- indirect governmental influence (for instance, subsidies on a new environmental technology).

- plan The activity is made up by a coherent set of relatively less precisely defined sub-activities with a joint aim (for instance, a structure planned for physical planning).
- regulation: The activity compromises all measures that may have an indirect impact (mainly via related activities) on environmental quality (for instance, environmental standards, charges, subsidies).

Both plans and regulations have a sector nature or a facet nature. The spatial scale of activities may relate to:

- international activities (cross-boundary transportation, e.g.)
- national activities (environmental impact regulation, e.g.)
- regional activities (location of an industry, e.g.)
- local activities (urban traffic rules, e.g.)

By combining now the attributes characterizing the nature of activities with the spatial scale, one may construct an activities profile.

The effects

Effect analysis aims at assessing the foreseeable consequences of various activities. The following effects in the framework of environmental management are relevant:

A. environmental effects:
(1) soil and ground water
(2) surface water
(3) air
(4) plants

|  |  |  |
|---|---|---|
|  |  | (5) animals |
|  |  | (6) landscape |
|  |  | (7) noise annoyance |
|  |  | (8) climate |
| B. | non-environmental effects: | (9) employment |
|  |  | (10) income |
|  |  | (11) accessibility |
|  |  | (12) housing market |
|  |  | (13) energy use |
|  |  | (14) facilities |
|  |  | (15) security |
|  |  | (16) health |

The effects can also be classified according to their features:

|  |  |  |
|---|---|---|
| A. | temporal effects : | (1) unique |
|  |  | (2) repetitive |
|  |  | (3) continuous short-term |
|  |  | (4) continuous long-term |
| B. | spatial aspects | (5) stationary |
|  |  | (6) mobile |
|  |  | (7) international |
|  |  | (8) national |
|  |  | (9) regional |
|  |  | (10) local |
| C. | remaining features : | (11) formal regulations applicable |
|  |  | (12) formal regulations not applicable |
|  |  | (13) marginal impact of effects |
|  |  | (14) non-marginal impact of effects |

It is clear that - as the next step of the typological approach - an impact (or effect) table can be constructed which comprises all activities and their expected consequences.

The decision structure

The decision problem in the framework of environmental management may have the following characteristics:

| | | |
|---|---|---|
| A. | Alternatives : | discrete number<br>continuous |
| | | point alternatives<br>sequential alternatives |
| | | mutually exclusive alternatives<br>mutually non-exclusive alternatives |
| B. | Information content : | quantitative<br>qualitative |
| | | certain<br>uncertain |
| | | extensive<br>limited |
| | | complete<br>incomplete |

The decision space of environmental management problems can be characterized by:

| | | |
|---|---|---|
| A. | The institutional : structure | international<br>national<br>regional<br>local |
| | | single objective<br>multiple objectives |

hierarchical
negotiation
informal

routine
non-routine

analytical
heuristic

B. The aim of the evaluation : ex ante evaluation
ex post evaluation

{ internal communication
{ external communication

identification of one alternative
{ identification of feasible alternatives
ranking of all alternatives

Clearly, the above-mentioned features can also be included in a matrix that combines activities and characteristics of the decision problem at hand, but for the sake of brevity this matrix will not be presented here.

Next, the features of environmental problems listed above can be translated into explicit criteria for selecting evaluation methods. By comparing these criteria with features of available methods insight is obtained in the relative usefulness of different methods for different problems.

Selection criteria are divided into <u>first and second order criteria</u>:

- First order criteria are mandatory binary criteria for the selection of an evaluation method; if a method does not comply with all first order criteria which are relevant to a certain problem this method cannot be applied to this problem.

- Second order criteria are not a priori mandatory criteria for the selection of an evaluation method. One tries to find a method which complies with as many second order criteria as possible but only a few, depending on the actual circumstances, will in practice function as mandatory criteria.

Part of the selection criteria are relevant to all evaluation problems.

If one designs for a specific field of policy the above-mentioned series of tables reflecting the features of the decision problem at hand, one may by means of a process of elimination and choice indentify the most appropriate method(s) for the specific decision situation concerned.

## PART F : BEYOND ASSESSMENT/MONITORING AND MANAGEMENT

F1 : Canter, L.W. and Fairchild, D.M.,
Posteis Environmental Monitoring

F2 : Carley, M.J.,
From Assessment to Monitoring: making our activities relevant to the policy process

# POST-EIS ENVIRONMENTAL MONITORING

L.W. Canter*
D.M. Fairchild**

## ABSTRACT

The paper deals with four P´s related to post-EIS environmental monitoring: purposes, purpos premises, planning considerations and potential p blem issues. Environmental monitoring programs carried out following the completion of environmer impact studies and the preparation of environmer impact statements (EISs). Environmental monitorin emerging due to the growing realization of the im tance of environmental monitoring over the life cy of a project that has environmental monitoring the life cycle of a project that has environmer impact.

## KEY WORDS

Environmental monitoring, environmental impact dies, environmental impact statements.

---

* Environmental and Ground Water Institute Univer of Oklahoma
** Environmental and Ground Water Institute, Unive ty of Oklahoma

Environmental monitoring refers to the set of activities which provide chemical, physical, geological, biological, and other environmental data required by environmental managers (U.S. Environmental Protection Agency, 1985). Due to the growing realization of the importance of environmental monitoring over the life cycle of a project, increasing attention is being given to the planning and implementation of monitoring programs following the completion of environmental impact studies and the preparation of environmental impact statements (EISs). Components within the broad definition of environmental monitoring include: planning the collection of environmental data to meet specific objectives and environmental information needs; designing monitoring systems and studies; selecting sampling sites; collecting and handling samples; laboratory analyses; reporting and storing the data; assuring the quality of the data; and analyzing, interpreting, and making the data available for use in decision making (U.S. Environment Protection Agency, 1985). Recognizing these components, the purpose of this paper is to explore four P´s related to post-EIS environmental monitoring: purposes; premises; planning considerations; and potential problem issues.

## Purposes of post-EIS environmental monitoring

The current Council on Environmental Quality (CEQ) regulations related to implementing the procedural provisions of the National Environmental Policy Act (NEPA) enunciate the principle of post-EIS environmental monitoring in sections 1505.3 and 1505.2(c) as follows (Council on Environmental Quality, 1978):

> Section 1505.3 --
> Agencies may provide for monitoring to assure that their decisions are carried out and should do so in important cases. Mitigation and other conditions established in the environmental impact statement or during its review and com-

mitted as part of the decision shall be implemented by the lead agency or other appropriate consenting agency. The lead agency shall:

(a) Include appropriate conditions in grants, permits, or other approvals.
(b) Condition funding of actions on mitigation.
(c) Upon request, inform cooperating or commenting agencies on progress in carrying out mitigation measures which they have proposed and which were adopted by the agency making the decision.
(d) Upon request, make available to the public the results of relevant monitoring.

Section 1505.2(c) --
At the time of its decision or, if appropriate, its recommendation to Congress, each agency shall prepare a concise public record of decision. The record, which may be integrated into any other record prepared by the agency, shall state whether all practicable means to avoid or minimize environmental harm from the alternative selected have been adopted, and if not, why they were not. A monitoring and enforcement program shall be adopted and summarized where applicable for any mitigation.

It should be noted that the CEQ regulations primarily focus on monitoring in conjunction with mitigation measures. Mitigation includes avoiding the impact altogether by not taking a certain action or parts of an action; minimizing impacts by limiting the degree of magnitude of the action and its implementation; rectifying the impact by repairing, rehabilitating, or restoring the affected environment; reducing or eliminating the impact over time by preservation and maintenance operations during the life of the action; and/or compensating for the impact by replacing or providing substitute resources or environments (Council on Environmental Quality, 1978). Monitoring could be used to determine the effectiveness of each of the types of mitigation measures.

Marcus (1979) identified the following six purposes or uses of information from the conduction of post-EIS monitoring:

(1) Provides information for documentation of the impacts that result from a proposed Federal action, with this information enabling the more accurate prediction of impacts associated with similar Federal actions.

(2) The monitoring system could warn agencies of unanticipated adverse impacts or sudden changes in impact trends.

(3) The monitoring system could provide an immediate warning whenever a pre-selected impact indicator approaches a preselected critical level.

(4) Provides information which could be used by agencies to control the timing, location, and level of impacts of a project. Control measures would involve preliminary planning as well as the possible implementation of regulation and enforcement measures. If an intergovernmental monitoring system is used it would facilitate appropriate response measures.

(5) Provides information which could be used for evaluating the effectiveness of implemented mitigation measures.

(6) Provides information which could be used to verify predicted impacts and thus validate impact prediction techniques. Based on these findings the techniques; e.g., mathematical models, could be modified or adjusted as appropriate.

## Premises basic to Post-EIS monitoring

Since post-EIS monitoring is becoming increasingly used as a component of continuing environmental management, and since effective monitoring programs require careful planning, it is useful to consider several premises basic to the required detailed planning. These premises, not necessarily in their order of importance, are:

(1) There is an abundance of environmental monitoring data routinely collected by various governmental agencies and the private sector. This data needs to be identified, aggregated and interpreted.

(2) Environmental monitoring programs are expensive to plan and implement; therefore, every effort should be made to utilize extant monitoring programs or modify extant programs as appropriate.

(3) Due to overlapping environmental management and monitoring responsibilities of many local, state, and Federal governmental agencies, it will be necessary to carefully coordinate environmental monitoring planning with several agencies.

(4) A basic need in previous, current, and future environmental monitoring programs is for scientifically-based interpretation of the collected information. All too often, information is compiled but never interpreted relative to the quality of the environment being monitored.

(5) Irrespective of the level of effort in an environmental monitoring program, there will never be enough information collected to thoroughly answer all questions which might be posed in a post-EIS monitoring program. Therefore, it will always be necessary to extend monitoring data via professional judgement to enable the making of an "educated best-guess".

## Planning considerations

Environmental monitoring systems can be considered in two phases: (1) development of a monitoring system; and (2) implementation and operation of a monitoring system. Planning is required for both phases. Figure 1 identifies eleven work elements associated with development of a monitoring system, and Table 1 provides brief explanations of the objectives and tasks associated with each element (Marcus, 1979). Key points to note from Figure 1 and Table 1 are:

(1) Work elements 1 through 3 are related to the preparation of an EIS.

(2) The monitoring objectives should be related to the anticipated impacts of the action. Impacts can occur on the physical-chemical, biological, cultural and socio-economic components of the environment.

(3) Coordination with the on-going monitoring programs of various agencies is vital in the development of a post-EIS monitoring system.

(4) Work element 7 on determining data requirements is the key technical element.

(5) It is important to utilize the data routinely collected under existing governmental programs.

(6) It will be necessary to adjust the post-EIS monitoring program to coincide with available budgetary resources. Several iterations may be necessary to achieve a workable monitoring system.

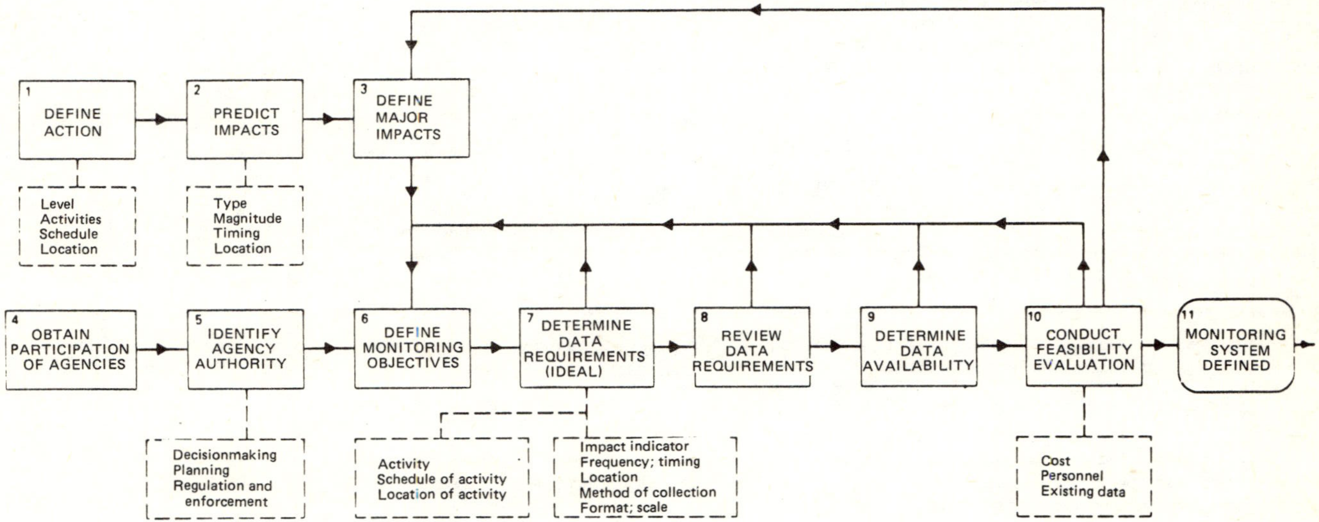


Figure 1: Monitoring Methodology Flow Chart: Phase 1 -- Development of a Monitoring System. (Flow chart should be used with explanation provided in Table 1. The solid numbered boxes correspond to the work elements listed in the table. Dashed boxes are annotations. The rounded box represents a decision point.)

Table 1: Explanation of Flow Chart of Monitoring Methodology: Phase I -- Development of a Monitoring System

| Work Element | Objectives | Tasks Necessary to Achieve Objectives |
|---|---|---|
| 1. Define action | To determine type, scope, level, location, and timing of action(s) causing impacts. | 1. Describe action by its activity components.<br>2. Determine activities schedule. (Timing and duration of activities affects timing of impacts.) |
| 2. Predict impacts | To determine type, magnitude, timing, and location of impacts resulting from the action. | 1. Prepare environmental impact statement. (EIS identifies impacts associated with each aspect of the environment and the activities causing the impact.) Predict impact magnitude, timing, and area to be affected. |
| 3. Identify and define major impact | To identify impacts that should be monitored. | 1. Identify impacts to be monitored on the basis of impact significance described in the EIS. |
| 4. Obtain participation of agencies | To obtain participation of agencies in development of a monitoring system. | 1. Contact all agencies having pertinent program responsibilities in area to be impacted.<br>2. Set up interagency meetings. |
| 5. Identify agencies' authority for controlling or mitigating impacts | To identify agency potential for responding to impacts through their decision-making, planning, regulation, and enforcement authority. | 1. Identify agencies' major areas of environmental concern. Determine for what aspects of the environment and for what type of impacts the agencies are responsible.<br>2. Identify individual agency basis of authority to control impacts through decision-making, planning, regulation, and enforcement. |

Table 1: (continued)

| Work Element | Objectives | Tasks Necessary to Achieve Objectives |
|---|---|---|
| 6. Define monitoring objectives | To define monitoring objectives. | 1. Define monitoring objectives in terms of major potential impacts and in terms of agency authority. |
| 7. Determine data requirements | To determine data needs for achieving monitoring objectives. (Reiterations of this element may be necessary, based on revision of monitoring objectives, testing, and evaluation of data usefulness. See feedback loops on Figure 1.) | 1. Reevaluate impacts on the basis of monitoring objectives; eliminate overlap in monitoring objectives and monitoring effort.<br>2. Select impact indicators. (These are the parameters that must be monitored to assess the magnitude of impacts. Several parameters may be indicative of a particular impact. Any impact indicator should be selected on the basis of its utility for decision-making, planning, regulation and enforcement.)<br>3. Determine frequency and timing of data collection. (Frequency of data collection should be the minimum necessary for trend analysis, enforcement of regulations, and correlation of cause and effects. For some parameters the timing of data collection may be more important than the frequency level; for example, collection of water-quality data during a major runoff event is more important than a precise data collection frequency. Timing of data collection should relate to the timing of activities causing the impact. Different phases of an action may produce different impacts that persist after an activity ceases.)<br>4. Determine monitoring sites or collection areas. (These should be based on the location |

Table 1: (continued)

| Work Element | Objectives | Tasks Necessary to Achieve Objectives |
|---|---|---|
| | | of the activities causing the impacts, predictions of areas most likely to be affected, and locations where integrated measurements would assist in gaining comprehensive understanding.) |
| | | 5. Determine method of data collection. |
| | | 6. Determine data type and storage format. (Data format possibilities include statistical tables, charts, graphs, summaries, maps, map overlays, computer printouts, and graphics. Criteria for selecting suitable format include: easy and convenient access to data by all users, intelligibility, interrelatability among formats, and ease of updating.) |
| | | 7. Determine data analysis method. |
| 8. Review data requirements in terms of monitoring objectives | To ascertain that data needs defined in work element 7 will fulfill monitoring objectives. | 1. Review data needs for conformance with monitoring objectives. |
| | | 2. Revise data needs as necessary to meet monitoring objectives. |
| 9. Determine data availability | To determine extent to which data needs can be met under existing agency programs, and to define data gaps. | 1. Identify what requirements are being met under existing programs, including frequency and timing of data collection, data collection location, accuracy, and method of collection. |

Table 1: (continued)

| Work Element | Objectives | Tasks Necessary to Achieve Objectives |
|---|---|---|
| 10. Conduct feasibility evaluation | To determine if proposed monitoring system is feasible within institutional, cost, funding, personnel, and time constraints. | 1. Determine cost, personnel, and time requirements for obtaining data.<br>2. Determine agencies' capability for providing data.<br>3. Determine whether proposed monitoring system is feasible. If so, work element 11 has been achieved. If not, work elements 1 through 10 must be repeated at a reduced level. (There are several potential approaches for reducing the monitoring system to a feasible level: the scope of monitoring objectives can be reduced; alternative impact indicators can be selected; the frequency of data collection can be reduced; and alternative methods of data collection can be used.) |
| 11. Define monitoring system. (This is a decision point. At this point participating agencies have decided what objectives the monitoring system will achieve, what impacts will be monitored, and what data will be collected. The next phase is implementation and operation of the monitoring system.) | | |

Several systematic approaches are available for allocating budgetary resources in a post-EIS monitoring program. One technique which can be used is the weighted rankings technique (Dean and Nishry, 1965). To illustrate the application of the technique to the allocation of budgetary resources, a hypothetical example problem will be presented. It is realized that this example is simplified; however, it is shown for purposes of illustrating the general procedure.

Assume that an EIS has led to the identificatin of five factors (indicators) considered important for inclusion in a post-EIS monitoring program. These factors are shown in this analysis as factors F-1, F-2, F-3, F-4 and F-5. The planning team also determined that the information for factors F-1 and F-5 was being routinely collected by other agencies. The only effort required to describe the environment relative to these two factors would be to procure the information and summarize it for inclusion in annual monitoring reports. The planning team further determined that there was some information being collected for factors F-2 and F-4, although it would have to be supplemented by short-term, site-specific, field studies. Finally, the planning team found that no information was available on factor F-3, and fairly extensive field studies would have to be accomplished to gather sufficient information for the environmental monitoring program.

The planning team, following considerable discussion, concluded that the five factors were not of equal importance in terms of the overall impact monitoring program. The problem then became one with a two-fold nature in that allocation of budgetary resources must be made between (1) five factors of unequal importance and (2) having differing quantities of information routinely collected (or requiring differing levels of effect to gather the necessary information). At this point the planning team decided to use a weighting technique which would aid in identifying the differing importance values for the five factors.

The weighting technique consists of considering

Table 2: Application of the Weighted-Rankings Technique to the Allocation of Budgetary Resources

| Factor | Assignment of Weights | | | | | | | | | | | | | | | Sum | FIC | Relative Cost for Data Collection | FIC X Relative Cost | Cost-Weighted FIC* |
|---|---|---|---|---|---|---|---|---|---|---|---|---|---|---|---|---|---|---|---|---|
| F-1 | 0 | 1 | 0 | 0 | 1 | | | | | | | | | | | 2.0 | 0.13 | 1 | 0.13 | 0.07 |
| F-2 | 1 | | | | | 1 | 1 | 1 | 1 | | | | | | | 5.0 | 0.33 | 2 | 0.66 | 0.31 |
| F-3 | | 0 | | | | 0 | | | | 1 | 1 | 1 | | | | 3.0 | 0.20 | 4 | 0.80 | 0.38 |
| F-4 | | | 1 | | | | 0 | | | 0 | | | 0.5 | 1 | | 2.5 | 0.17 | 2 | 0.34 | 0.16 |
| F-5 | | | | 1 | | | | 0 | | | 0 | | 0.5 | | 1 | 2.5 | 0.17 | 1 | 0.17 | 0.08 |
| F-6 (Dummy) | | | | | 0 | | | | 0 | | | 0 | | 0 | 0 | 0 | 0 | 0 | 0 | 0 |
| | Total | | | | | | | | | | | | | | | 15.0 | 1.00 | | 2.10 | 1.00 |

*These decimal fractions can be used in the proportionate allocation of budgetary resources.

each factor relative to every other factor and assigning to the one considered to be the most important of the pair a value of 1, and to the lesser important of the pair a value of 0 (Dean and Nishry, 1965). The use of this technique is shown in Table 2. In addition to the five basic factors a sixth factor, called a dummy factor, is also shown. The dummy factor is included so as to preclude the net assignment of a value of 0 to any one basic factor in the process of each paired comparison. The dummy factor is defined as the least important in every comparison. If two factors are considered to be of the same importance (one is not more important than the other), then a value of 0.5 is assigned to each of the pair. Following the assignment of relative weights, with this process only being completed following several iterations so as to be sure that the planning team is considering each factor in a consistent manner to each other factor, then the individual weight assignments are summed, with the factor important coefficient (FIC) being equal to the sum value for an individual factor divided by the sum for all of the factors. The total of the FIC column should equal to 1.00. The total of the sum column should equal to (N) (N-1)/2 where N is equal to the number of factors included in the assignment of weights. In this example, six factors were included, hence the sum total should be equal to 15.0.

The FIC column in Table 2 indicates that F-2 is the most important followed by F-3, F-4 and F-5, and F-1. If the data aggregation costs were the same for each of these factors, then the allocation of budgetary resources could be made based on the proportions shown in the FIC column. For example, if $50.000 was available on an annual basis for post-EIS monitoring, then $16.500 ($50.000 x 0.33) could be allocated for F-2, and so forth.

Recognition of the relative costs of the assemblage of information can lead to differing decisions on the allocation of budgetary resources. Table 2 also contains a column identifying relative costs for data collection. As noted earlier, factors F-1 and F-

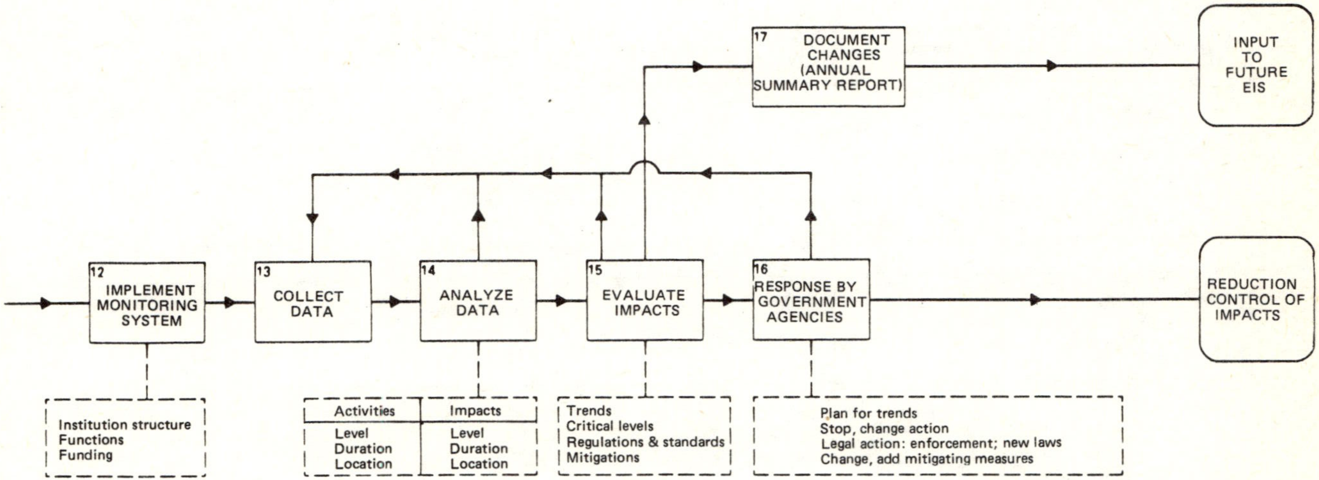


Figure 2: Monitoring Methodology Flow Chart: Phase II -- Implementation and Operation of a Monitoring System. (Flow chart should be used with explanation provided in Table 3. The solid numbered boxes correspond to the work elements listed in the table. Dashed boxes are annotations, The rounded boxes represent end points.)

Table 3: Explanation of Flow Chart of Monitoring Methodology: Phase II -- Implementation and Operation of a Monitoring System

| Work Element | Objectives | Tasks Necessary to Achieve Objectives |
|---|---|---|
| 12. Implement monitoring system | To implement the monitoring system. | 1. Review suitability of existing institutional entities for operation of monitoring system. |
| | | 2. Create institutional structure or modify existing institutional structure as needed. |
| | | 3. Define agency and institutional entity functions and responsibilities. |
| | | 4. Prepare formal written agreement of agency responsibilities; obtain approval. |
| | | 5. Obtain needed funding. |
| 13. Collect data | To collect data required for defined monitoring objectives. | 1. Collect data and submit results. |
| 14. Analyze data | To determine the level, location, and duration of activities and impacts. | 1. Determine activity and impact levels. |
| | | 2. Define location of activities and impacts. |
| | | 3. Determine duration of activities and impacts. |
| | | 4. Correlate activity and impact data. |
| 15. Evaluate impacts | To evaluate the significance of impact levels. | 1. Identify impact trends; identify rate of change. (The rate at which an impact is increasing is significant because of the need to respond to impact trends in a timely fashion before critical impact levels are reached.) |

Table 3: (continued)

| Work Element | Objectives | Tasks Necessary to Achieve Objectives |
|---|---|---|
| | | 2. Identify impacts that have reached critical impact levels. (Critical impact levels requiring immediate notification of participants should be set for each impact being monitored.) |
| | | 3. Identify impacts that have exceeded legal limits. |
| | | 4. Evaluate effectiveness of mitigating measures. |
| 16. Response by regulatory or responsible agencies | To use agencies' decision-making, planning, regulation, and enforcement authorities to reduce and prevent impact. | 1. Plan responses to impact trends. |
| | | 2. Respond to critical impact levels: stop or modify activities causing impact; treat impact. |
| | | 3. Respond to nonconformance with regulations and standards through enforcement and prosecution. Develop additional regulations, standards, and legal authority as needed. |
| | | 4. Respond to evaluations of mitigating measures: revise, terminate, or add mitigating measures as appropriate. |
| 17. Document changes | To maintain record of (1) impact levels that result from projects and (2) effectiveness of mitigating measures in reducing impact levels. | 1. Prepare annual summary report. |

5 will acquire minimal efforts, while factor F-3 will require a fairly extensive program. The relative cost column was developed by the planning team following discussion, with the relative costs including expenditures for necessary environmental sampling, analysis, interpretation, and assemblage of information. If the FIC-values are multiplied by the relative costs and those numerical values summed and proportioned, the cost-weighted FIC column is obtained. In this example, the cost-weighted FIC is 0.31 for F-2, 0.38 for F-3, and so forth. Based upon consideration of relative costs in conjunction with relative importance, the allocation of the \$50.000 would be \$3.500 to F-1, \$15.500 to F-2, \$19.000 to F-3, \$8.000 to F-4 and \$4.000 to F-5.

This example has illustrated how a planning team can allocate budgetary resources. These types of decision are made in every post-EIS monitoring program, whether specifically identified or not. Most programs are characterized by the allocation of budgetary resources pretty much on an equivalent basis, or an as-needed basis in accordance with a first-come, first-serve concept of expenditure. The particular approach discussed herein gives order and format to the decision-making associated with post-EIS monitoring planning.

The second phase of a post-EIS monitoring program involves the implementation and operation of the monitoring system. Figure 2 deineates the work elements, and Table 3 provides descriptions of the objectives and necessary tasks for each element (Marcus, 1979). Key points to note from this phase are:

(1) Work element 12 on implementing the monitoring system may require considerable effort in obtaining inter-agency agreements and necessary funding.

(2) Evaluation of impacts will involve the predetermination of criteria to be used for interpretation. These criteria should be based on legal limits, professional judgment and public

inputs.

(3) Development of appropriate response plans to impact trends can be time-consuming and technically difficult, and require considerable coordinating efforts.

(4) It is vitally important that annual summary reports, or reports on more frequent intervals, be prepared to document the findings and resultant responses of the post-EIS monitoring program.

Problem issues.

A number of problem issues can arise in conjunction with the planning and implementation of a post-EIS monitoring program. Examples are:

(1) There may be lack of quality control on data retrieved from computerized information storage and retrieval systems. To serve as an illustration, a water quality data base indicated that the dissolved oxygen in a flowing stream had been measured as 120 mg/l. This represents a physical impossibility under normal environmental conditions, and would suggest that a decimal point error (120.0 rather than 12.0) had been introduced into the data as it was being formatted for the information system.

(2) It may be difficult to integrate information from different sources for a particular study. The information may not be compatible and it can actually be contradictory. Even from the same source of information there can be inconsistencies in technical terminology and level of detail. Recent experiences have indicated that routine items such as station numbering systems and the identification of monitoring stations on maps is often lacking (Fairchild, Kamat and

Canter, 1982).

(3) A particular problem associated with environmental data is the general lack of data interpretation. Numerous agencies are proficient in monitoring system design and data collection; however they give less emphasis to data interpretation. Data interpretation is more important than data presentations. Accordingly, it may be necessary for the post-EIS monitoring team to interpret published data as part of their work.

(4) Several potential problems are associated with the growing usage of computer systems for source material identification and procurement. Many systems have been established by computer specialists and not by technical specialists in substantive areas. One potential problem is associated with the ability of technical specialists to communicate with computer specialists in a language mutually beneficial to both. Planning of information systems should be achieved through joint efforts, and system usage should be by persons who have technical training. It is difficult for non-trained technicians to extract the maximum amount of information from many computer systems.

## SUMMARY

Post-EIS monitoring is increasing as a result of CEQ requirements relative to mitigation measures. This component of environmental management over the life cycle of projects, particularly large-scale-projects, is expected to further increase. Systematic approaches exist for the design and implementation of post-EIS monitoring programs. While these approaches are useful, considerable flexibility is needed as more experience is gained on post-EIS monitoring. Many problem issues exist, and these must be systematically resolved if the overall goals of post-EIS monitoring are to be attained.

## SELECTED REFERENCES

Council on Environmental Quality, "National Environmental Policy Act -- Regulations", Federal Register, Vol.43, No. 230, Nov. 29, 1978, pp. 55978 - 56007.

Dean, B.V. and J.J. Nishry, "Scoring and Profitability Models for Evaluating and Selecting Engineering Products", Journal Operations Research Society of America, Vol. 13, No. 4, July-Aug. 1965, pp. 550-569.

Fairchild, D.M., R.P. Kamat and L.W. Canter, "Environmental Data for Assessment of Water Issues for Enhanced Oil Recovery", 1982, University of Oklahoma, Norman, Oklahoma.

Marcus, L.G., "A Methodology for Post-EIS (Environmental Impact Statement) Monitoring", Geological Survey Circular 782, 1979, U.S. Geological Survey, Washington, D.C.

U.S. Environmental Protection Agency, "Resource Document for the Ground Water Monitoring Strategy Workshop", March 1985, Office of Ground Water Protection, Washington, D.C.

## FROM ASSESSMENT TO MONITORING: MAKING OUR ACTIVITIES RELEVANT TO THE POLICY PROCESS

Michael J. Carley*

ABSTRACT

It is increasingly obvious that the static impact assessment process we have developed to date is insufficient in concept and practice to deal with the reality of a dynamic social and policy context. This paper argues that recent emphasis on socio-economic and environmental monitoring is therefore a logical, timely extension of the assessment process, and one which brings it more in line with the requirements of policy-making. The paper distinguishes assessment and monitoring. It then attempts to overcome confused perceptions between monitoring as a technical exercise versus monitoring as a strategic management and planning function by describing eight categories of environmental and socio-economic monitoring ranging from technical inspection to cumulative regional monitoring. The importance of understanding the organisational context of monitoring is stressed, and different audiences for monitoring are described. Finally the author draws on his experience in developing a regional monitoring programme in Canada to suggest essential criteria for the development of monitoring useful to the policy making process.

KEY WORDS

Socio-economic and environmental monitoring, assessment, monitoring, strategic management and planning function, technical inspection, cumulative regional monitoring

* Policy Studies Institute, 100 Park Village East, London NW1 3SR, United Kingdom

What is a socio-economic monitoring programme?

There is no ready, simple definition of a socio-economic monitoring programme. The definition of the word "monitor" itself differs according to source. The Oxford Concise Dictionary puts a monitor as "one who keeps order" while Webster defines monitor as "to observe or watch". The difference between these definitions, that is a monitor as a policeman ensuring good behaviour versus as a neutral observer, reflects a common distinction between different types of monitoring.

A useful beginning definition of monitoring is "the systematic collection of data comprising key indicators of social, environmental, and economic impacts" (Bankes and Thompson, 1980, p. 10). Certainly, the concept of monitoring is a logical outgrowth of the move to social and environmental studies, and monitoring has been called continuous assessment (Carley and Bustelo, 1984). There is a distinction, however, between social impact assessment and monitoring: the former anticipates future impacts based on experience in other similar situations and extrapolation of data, while the latter observes or reports on actual events, measured and analysed as they occur. The collection of data, or information gathering, is of no special use in itself, however, and here is where we must expand our definition of monitoring. Monitoring is the systematic collection and organisation of information, which is to be used in improving the decision-making process; either indirectly by informing the public, or directly as a feedback tool designed for purposes of project management, programme evaluation or policy development. Monitoring assists a government agency or private corporation in keeping its policies and decisions responsive to new opportunities and unforeseen changes in the decision environment.

Monitoring therefore can be viewed as an important component of the cyclical planning, program design, implementation, evaluation and replanning process, and can serve one or more of three organisational functions:

(1) Monitoring can enlighten us on the status of critical or changing issues in the policy environment which may need to be addressed by future action.
(2) Monitoring can help us manage our environment by providing feedback in terms of the relative success or failure of previous action in terms of policies and programmes.
(3) Monitoring can be used to test compliance with government regulations, and contractual agreements between government and industry.

So monitoring is a component of an organisational process, or more simply, a tool of organisation, be they government, industry, or community development groups. A recent report put community monitoring in an organisational framework (Procter and Redfern, 1979, p. 3.3):

> Any organisation includes to some degree the functions of objective setting, plan preparation and decision making. The organisation will also have feedback from these actions to the other functions, to indicate changes in the environment of the organisation, and the success and implications of the implementing actions and their outcome. Thus monitoring is an essential component of a continuing process and provides a coordinating and integrating function. It includes the collection of information, and also the analysis and evaluation of that information, and the provision of advice based on the evaluation.

A simple diagram shows the organisational context of monitoring:

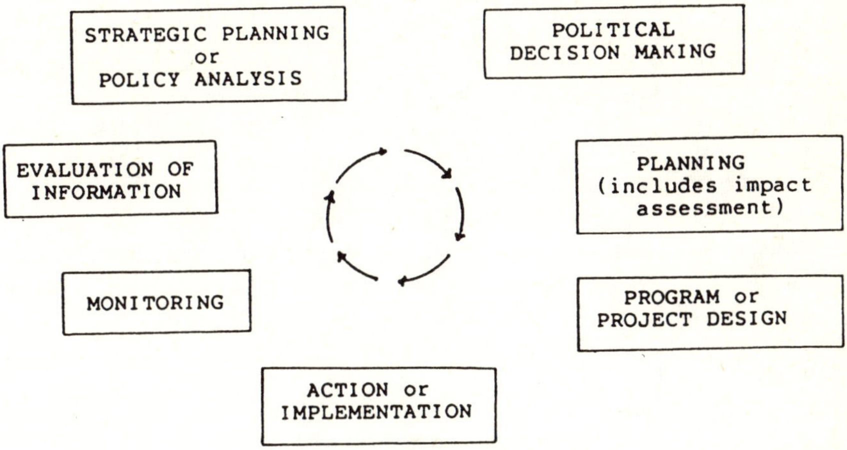


Figure 1

The important things to note about this diagram are:

(i) that it is circular, that is, it represents what is a continuing process of any number of cycles which result in policy and/or action,
(ii) that in some cycles some steps are left out, and
(iii) that monitoring does not occur in a vacuum but is part and parcel of a decision making/planning process.

But this general view of monitoring leaves us with some questions: are there different types of monitoring?; are there different ways of structuring a monitoring program?; and finally, who are the users of socio-economic impact monitoring? The next sections look at these issues.

## Eight types of monitoring

Monitoring is a term which has been bandied about of late, but there is generally some confusion among different people about what constitutes monitoring. Perhaps Figure 2.0 can help clear up some of the confusion. Given our very simple model of monitoring as a feedback process, it can then be applied at different geographic scales, and with different organisational approaches. There are at least eight types of monitoring which exist as kind of continuum of activity:

Inspection - the simplest form of monitoring is the periodic inspection of some activity for any stated purpose, for example, to ensure that safe operating procedures are being followed, and that visible environmental degradation is not occurring. For example, the Canadian Oil and Gas Lands Administration conducts regular on-site, bi-weekly inspections of oil and gas drilling activities by energy companies.

Regulatory permit monitoring - is the continual, or periodic, recording of industry performance related to environmental terms and conditions made pursuant to the insurance of exploration, construction, operation and pollution control permits, for assessing compliance and requests for permit renewals. For example, discharges from an industrial process into a stream are usually monitored. Inspection and regulatory permit monitoring are directed to human activities (for example, discharges), rather than to the bio-physical environment per se (for example, the stream itself).

Experimental environmental monitoring - is a scientific process designed to test specific

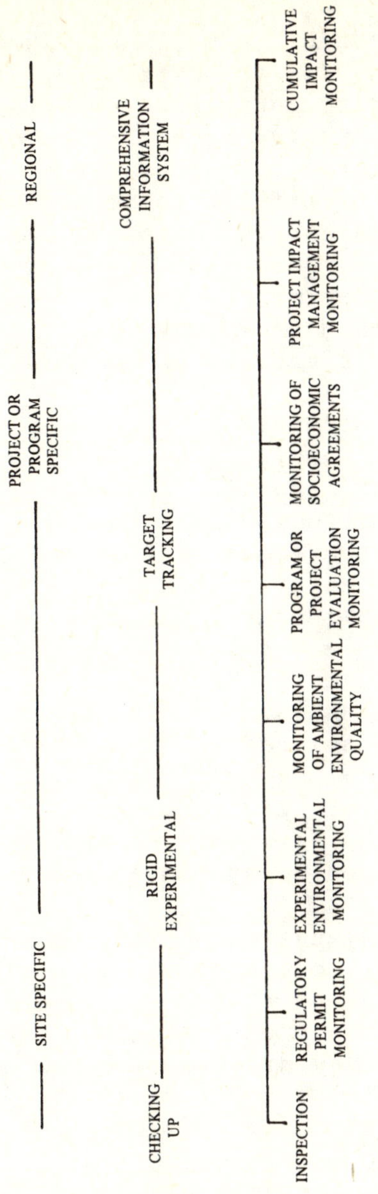


FIGURE 2.0 : MONITORING TYPES

hypotheses and predictions on how environmental change is caused and expressed. This monitoring is intended to contribute to scientific knowledge by an iterative accumulation of information, to update current concepts of ecological processes, and to provide background information on the environment which may be useful for project evaluation. There has also been social monitoring in experimental or quasi-experimental frameworks, but these have been generally unsuccessful for a variety of reasons.

<u>Monitoring of ambient environmental quality/environmental auditing</u> - is the repetitive measurement of specific ecological phenomena to document environmental change due to the direct or indirect effects of resource development or industrial activity, and to distinguish those changes from natural variability. Such monitoring is project-related, and may be used for purposes of testing the validity and reliability of impact assessment and testing mitigative measures (Beanlands and Dinker, 1983). The focus is on the environment itself. A Canadian example is the Beaufort Environmental Monitoring Project (1983) which is examining project-related changes to "valued ecosystem components".

<u>Programme evaluation monitoring</u> - also called productivity measurement or performance auditing. These are attempts, generally by government departments or evaluation teams, to measure the efficiency or effectiveness of policies or programmes in terms of the ratio of organisational inputs to outputs (efficiency) or satisfaction of goals and objectives (effectiveness). Such studies are designed to (i) help determine progress towards targets or goals set by public administrators, (ii) identify problem areas and help set priorities for efforts at improved productivity and better value-for-money in government expenditure, (iii) help implement wor-

ker incentive schemes. Such monitoring efforts are usually focussed on a single set of policies or a particular service delivery programme.

Project evaluation monitoring - similar to the above and most common in developing countries where social development projects have been funded by international donor agencies such as the Canadian International Development Agency, the World Bank, and others. Also called project audits, or post-project audits, these are attempts to determine the relative success or failure of aid projects while they are ongoing or completed, not only in cost-benefit economic terms, but in social and environmental terms as well. The focus again is on evaluation of efficiency and effectiveness, and on value-for-money in relatively discrete projects. Both programme and project evaluation are part of what is called "evaluation research" and there is crossover between these activities and impact monitoring. In some cases, however, similar terminology is used, but with some different meanings.

Monitoring of socio-economic agreements or contracts - this is the monitoring of voluntary or contractual agreements between industry and government. Such monitoring focuses on the performance of industry in terms of industrial benefits, employment and training benefits, social and cultural expenditure, and public consultations. In a general sense such monitoring can be orientated towards provision of specific benefits, or the mitigation of specific costs or negative conditions, or both.

Monitoring for project impact management - this is a broader form of monitoring in which a formalised organisation or group regularly assesses a wide range of environmental, social, and economic impacts from a particular project or activity. These may range from resource development and construction projects to industrial plant closure. This monitoring is designed to ensure a flow of benefits, or to control the severity and distribution of anticipated or especially unanticipated negative effects as they occur, and possibly to ensure due compensation where it is required. This type of monitoring is directly linked to schemes to reduce or mitigate negative impacts. Organisationally, at its simplest it is carried out by project proponents and overseen by an appropriate government agency. More realistically, it is conceived of as high-level monitoring by responsible public agencies in association with independent groups. The intent is to report on the entire project on activity, including objectives; criteria for judging success; the technical and institutional means for reaching these; planning systems; adequacy of inspections and coordinations; and other critical social issues like unemployment (Nelson and Jessen, 1981). This is an important extension of the impact assessment process. It attempts to overcome deficiencies in that process, and especially with predictive impact statements. Such statements are static rather than dynamic in nature, and therefore unresponsive to new information and changing conditions. A side benefit of monitoring at this scale may be ex ante testing of predictions in the original social impact assessment, similar to environmental audits.

Cumulative impact monitoring - this is usually of interest in geographic regions where rapid industrial or agricultural development and/or dramatic socio-economic change is taking place, as well as in areas of deindustrialisation. This monitoring programme is not confined or limited to programme or project-related direct or indirect impacts. Rather the focus is on monitoring all critical issues or changing socio-economic patterns in a region, whether they are related to a project or government programme or occur independently. The cumulative perspective for human environment considers the social, economic, cultural and political implications of all structural economic changes in a city or region with attention to their interrelated effects over time. The sum of these interrelated effects is likely to be greater than those generated by particular policies, programmes, or projects considered seperately. That is, impacts may have an additive effect: many small, local impacts could be serious when occurring together. Such monitoring is characterised by a area-wide or regional, rather than site-specific, perspective; attention to overlapping impacts of different projects and policies; and an time-perspective stressing the long term, incremental and dynamic nature of social change. It is usually carried out by government to (i) provide a regional overview, (ii) to provide a coordinated and organized flow of information for strategic planning purposes, and (iii) to ensure some compatability in the impact of different policies eminating from different government departments. In addition to documenting socio-economic change, it also serves to coordinate the variety of information generated by less strategic types of monitoring. It can begin well before decisions on project acceptability are taken, but more likely it will be a component of a regional or strategic planning process, and therefore unrelated to the timing or acceptability of particular projects. The cumulative monitoring approach is also of interest in cases where cause and effect, of activity related to impact, is difficult to assess because of numerous economic and social interventions or because of a complex socio-economic environment (for example in most urban areas). The cumulative monitoring approach sidesteps the vexatious issue of causation in the social environment, by choosing to focus on the critical issues which confront the residents of an area under study. We will return to this point in a moment.

There are some interesting and varying dimensions to these different types of monitoring. The level of political contentiousness of the issues considered in the monitoring effort rises as the policy level moves from day-to-day management to the level of strategic decisions. Such strategic decisions are characterised by complex problems, a variety of alternative policies, broad decision criteria, and imprecise information. Geographically, monitoring efforts may be site-specific, programme specific, project specific, or regional, i.e. multi-project. Lastly, the approach may vary. At its simplest, monitoring is no more than formalised "checking-up". Experimental monitoring, on the other hand, demands a rigid scientific approach with careful isolation of independent from dependent variables. Many monitoring efforts involve the target-tracking approach, which establishes a target or forecast for a particular variable of interest, usually related to time. The variable may range from a pollutant level in regulatory environmental monitoring, to a proportion of locally-based new employees in a work force in benefit agreement monitoring. The monitoring system then tracks the variable at set time periods to identify deviation from the target. Finally, project impact and cumulative impact monitoring demand an information system approach which identifies a wide range of critical and interrelated variables to be monitored, both quantitative and non-quantitative, within some sort of organised data framework. The next sections look briefly at the audience for cumulative impact monitoring.

## Who are the users of monitoring?

A local citizens group in Canada in considering oil and gas related development recently argued:

> We are concerned about the lack of effort on the part of both the developer and the government to ensure that the people affected have a clear understanding of the nature and long-term consequences of the proposed development.

This suggests a first audience for monitoring: the public at large, and community and social development groups. This is especially true of cumulative impact monitoring which should be designed and published so as to enable all citizens of a region to better understand the social, economic and cultural changes occurring in their area, and the numerous policies and initiatives related to those changes. Such an information flow encourages participation in the pluralist, or democratic, process and may help to strengthen local control as an industrial culture confronts a formerly land-bound or agricultural one. The community is not a passive audience, but may take or encourage social action based on monitoring information.

The second audience for monitoring is the strategic policy maker who needs to identify particular issues requiring a policy resolution, and who uses the information from the monitoring programme as an agenda for political decision making, often between or among competing programmes or objectives. This broad policy making activity, whether at the departmental, or strategic inter-departmental, level also requires a sensitivity to the "interrelatedness" of decisions. The monitoring program helps an understanding of a broad range of issue areas which are interconnected. While sophisticated social modelling in a quantitative mode is impossible, it is part of the art of policy making to group the interconnected cause and effects of decision and action in various policy spheres. The operational aspects of such strategic (intergovernmental or interdepartmental) policy-making is anticipatory or forward looking planning, which attempts to anticipate future issues and structure goals and objectives accordingly. This is the opposite of ad hoc planning, or "firefighting", which deals with issues only when they become critical problems.

The third audience for monitoring information is the departmental programme planner who can use that information as a background briefing on the status of change in different geographic areas, in different government departments, and at different levels of government. Monitoring provides a useful context within which to consider more detailed programme evaluations.

The fourth audience for monitoring information is project proponents or industrial developers. They also will be directly involved and interested in less strategic forms of monitoring (especially of regulatory permits and socio-economic agreements), but as reasonable corporate citizens they may also be interested in cumulative impact monitoring. This will affect them directly, as they make self-adjustments to policy and operations; and indirectly, as government or public pressure suggests changes in policy or operation as required.

## A focus on issues in monitoring.

The main methodological problem in developing a monitoring or socio-economic indicator system is in deciding what needs to be measured: what to include and what to leave out of the monitoring report. Things omitted mean criticisms of ignoring or hiding information. If too much is included reports become unwieldy, unread, and contribute to information overload.

Past experiences at monitoring are not much help. In some cases the list of indicators reflects more or less what data are available and not much else. Such an approach may tend to ignore what is important in favour of what data exists. In other cases, the list of indicators represents not more than the opinion of the researchers on what is important. In still others a very comprehensive list is drawn up but with no attention to whether any data can be collected without prohibitively expensive survey research. Some monitoring efforts are based on good lists of indicators, usually where the objective of the programme is to measure progress towards specified planning objectives, for example, the monitoring of structure plans by some County Councils in the UK. However, this doesn't help in cases where planning objectives are not readily formulated.

Elsewhere the author (Carley, 1981, p. 157) has argued that no socio-economic indicator should be included in a social report unless it has been justified by a "measurement rationale". This is a written statement which explains why a statistic should be regarded as a measure of a social variable under consideration, and why that variable is to be considered important. By making the argument explicit, the researcher invites people to examine his logic and accept or reject it. Rejection will suggest the monitoring programme needs to be overhauled. Sufficient acceptance (not everyone will agree, no matter what) suggests that there is support for the monitoring programme as a window on reality. A measurement rationale also reduces the dangers of "cultural imperialism" in which the researcher's own values substitute for the values of those living in the community or region under study.

The measurement rationale is based on the following question: what are the important social, cultural and economic concerns facing the people of the geographic area under consideration? The selection of indicators is then derived from a thorough examination of these issues. In every case the presentation of an indicator is justified by linking that indicator with critical social issues. Issues are paramount, indicators help us think about issues. The issues themselves are often politically contentious, but the report tries not to take sides. It merely says: there is strong evidence that this is an issue which should be monitored to contribute to discussion on the social development of the community or region in question.

## Some lessons for monitoring

The following points are suggested by the author's recent review of monitoring efforts:

(1) Monitoring is an important element in the post-approval stage of project appraisal, but a proposed monitoring program should not play any major part in the determination of project acceptability on economic, social and environmental grounds. Monitoring may be a condition of project approval, which is based on more important factors. There may be a tendency to confuse grounds for project approval, with limiting conditions or constraints on the approval. Too often these days, the provision of monitoring is seen as somehow helping to justify projects. It can do no such thing and project acceptability must be based on substantive socio-economic issues, not on procedural grounds.

(2) The objectives of a monitoring programme cannot be assumed but must be clearly thought out. This is the first step in the development of such a programme. The objectives will differ for different types of monitoring. For example, a cumulative monitoring program may have a major objective of providing broad regional and community level information on scoio-economic changes and trends on perhaps an annual basis, reported to the public, government and industry. Project impact monitoring will have an important objective of identifying project-induced impacts and linking these to schemes to mitigate negative effects.

(3) The institutional roles of various individuals and committees in any monitoring programme should be well-defined, and designed specifically to fulfill the stated programme's objectives. The whole proposal should be as simple as prac-

ticable.

(4) The individuals and committees involved in monitoring must have sufficient authority to carry out the stated objectives. Such authority may have to be delegated from a variety of government sources and in this sense monitoring will require an act of political will and commitment on the part of the agencies involved. The monitoring group, on the other hand, should answer directly to one responsible coordinating committee which is the channel by which the entire monitoring programme relates to a variety of agencies at different levels of government, and to industry. Monitoring often requires a good measure of inter-agency cooperation, as important social and economic effects in peoples´ lives are not easily divided into convenient agency categories. Anyone who has thought about the links among primary health care, education, social development and income will recognise this.

(5) Monitoring is likely to highlight critical socio-economic issues and problems on the community, and thus raise political consciousness on those issues. If at all possible the monitoring exercise should be linked with community development or action programmes, as well as provide direct communication with social agencies with interest and resources aimed at ameliorating social problems, or assisting the community to do so.

(6) A monitoring programme with strong community links should be designed in such a way as to be an open process. The public must have confidence in the monitoring programme, and this will certainly not occur if it works _in camera_. The public should be able to readily ascertain who is responsible for what in the monitoring programme and how and where the public has access

to that programme, preferably through a community-based advisory committee. This may be less of an issue if monitoring is regional or national in scope.

(7) Monitoring should begin in the planning process and continue through to the operational phase of the project. The committees to which the monitoring programme reports must therefore have jurisdiction over time.

(8) Finally, methodological issues in monitoring design may be subservient to organisational issues, but they are not unimportant. A proper data base for monitoring socio-economic and environmental indicators is required, as is the "measurement rationale" for the selection of indicators.

## REFERENCES

Bankes, N. and A.R. Thompson (1980), Monitoring for Impact Assessmnt and Management, Vancouver: Westwater Research Centre.

Beanlands, G.E. and P.N. Duinker (1983), An Ecological Framework for EIA in Canada, Halifax: Institute for Resource and Environmental Studies, Dalhousie University.

Beaufort Sea Environmental Monitoring Program (1983), Preliminary Survey Report for Workshop II, Ottawa: Indian and Northern Affairs Canada.

Bisset, R. (1984), "Role of Monitoring and Auditing in EIA", Aberdeen: Centre for Environmental Management and Planning.

Carley, M.J. (1981), Social Measurement and Social Indicators: Issues of Policy and Theory, London and Boston: George Allen and Unwin.

Carley, M.J. (1984), Cumulative Socio-Economic Monitoring, Ottawa: Northern Economic Planning Branch, Indian and Northern Affairs Canada.

Carley, M.J. and E.S. Bustelo (1984), Social Impact Assessment and Monitoring: A Cross-disciplinary Guide, Boulder: Westview Press.

Nelson, J.G. and S. Jessen (1981), The Scottish and Alaskan Offshore Oil and Gas Experience, Ottawa: Canadian Arctic Resources Committee.

## PART G : IMPACT ASSESSMENT IN DEVELOPING COUNTRIES

# ENVIRONMENTAL ASSESSMENT AND DEVELOPMENT ASSISTANCE THE WORK OF THE OECD

F.W.R. Evers*

## ABSTRACT

The OECD *ad hoc* Group on "Environmental Assessment and Development Assistance" made up of representatives from environmental and development assistance ministries of OECD member states, was formed in October, 1983 to carry out a programme of work based on four objectives. These objectives are:

(1) To identify those types of development aid projects most in need of environmental assessment;
(2) To identify the constraints faced by developing countries in assessing the environmental consequences of development projects and how those constraints can be overcome;
(3) To examine the experience to date of OECD member states in actually assessing the environmental impacts of their development projects and programmes; and
(4) on the bases of above, to determine the types of procedures, processes, organization and resources needed to ensure that environmental assessment can be undertaken in a satisfactory, timely and cost-effective way.

Work on the first two objectives has been completed and a draft recommendation prepared which includes a list of those activities most in need of assessment. They include, among others, industrial activities, infrastructure projects and the exploitation of hydraulic resources. In addition, certain fragile environments were identified where, if a development is planned, in-depth assessment should be carried out regardless of the project type. Examples of such environments include wetlands, mangrove swamps and coral reefs.

In carrying out the second objective, the Group called upon the experience of member state aid and environment agencies. It also undertook an extensive literature search among a number of sources including UN bodies, other international organizations, universities, research institutes and of course, developing countries themselves. The resultant identification of constraints encluded such hinderances as lack of scientific data, institutional structures and skilled manpower as well as the need for political awareness. These, and other constraints, have been described along with specific examples of how they have been overcome.

In addressing objective 3 the group assembled sixteen case studies on development aid projects which had been subjected to some kind of environmental assessment. The case studies were submitted by the governments of The Netherlands, U.S.A., Canada, the United Kingdom, France, Denmark, Norway, Australia and Japan. They describe a number of projects, including highways, electrical power generation, pesticides, dams and waste management.

The case studies revealed a number of key elements for a successful environmental assessment process. Experience has shown, for example, that early "scoping" of environmental impacts can result in a simplified form of assessment that identifies major potential consequences as well as the design and/or mitigation measures needed to counter them.

The case study conclusions, along with the findings under the first two objectives, will form the basis for the Group´s final report which will be issued at the end of 1985.

KEY WORDS

Environmental impact assessment, development assistance, constraints, experience, procedures

---

* Deputy Director-General for Environment, Ministry of Housing, Physical Planning and the Environment

The Organisation for Economic Cooperation and Development - the OECD - has been giving attention to environmental impact assessment since 1974. That attention has been centered primarily in the Organisation´s Environment Committee where, as in other OECD committees, representatives of the 23 member countries meet to discuss common problems, the international economic implications of these problems and solutions to them. The Environment Committee has passed to recommendations related to EIA - one in 1974 and the other in 1979. The recommendations call upon to the Member countries to establish procedures and methodologies for assessing the environmental impacts of significant public and private projects and to exchange information on matters which could help them better forecast environmental effects.

Since the passage of those recommendations most OECD Member countries have indeed established some type of EIA system. No two countries, however, have taken an identical approach. As a result, the scope and requirements of national EIA processes vary greatly. Nonetheless, OECD Member countries are gaining experience in assessing the environmental impacts of their domestic activities - in both the public and private spheres. The application of EIA to development aid activities, however, has been much less common. There are probably several reasons for this - one main one being that aid policy has not been subject to the same internal political pressures as matters related to the domestic environment. But the situation is changing and the need to inject an environmental component into aid programmes is being recognized more and more.

This brings me to the OECD´s most recent efforts in the field of EIA and the subject of my remarks today - the work of the ad hoc group on Environmental Assessment and Development Assistance of which I am the chairman. Our group represents a cooperative effort between the Environment Committee and the Development Assistance Committee working together for the first time. The group which had its first meeting in October, 1983, is made up of representatives from

Member Countries environmental and development aid administrations.

At our first meeting we agreed upon a programme of work centered around four objectives. They are:

(1) To identify the kinds of development aid projects and programmes that are most in need of environmental impact assessment;
(2) To identify the constraints faced by developing countries in assessing the environmental impacts of aid projects and programmes;
(3) To examine the experience of aid agencies in actually carrying out assessments, and
(4) based on the first three objectives, to make recommendations regarding the kind of procedures, organization and resources needed to ensure that aid projects and programmes are adequately assessed for their environmental impact.

I would like to briefly go through these four objectives with you and point out what we have learned. Before doing that I should mention out that we are now nearing the completion of our work and should have the final report completed by the end of this year.

There are a number of reasons for identifying the types of aid projects and programmes "most in need of EIA" - our first objective. For example, some activities, because of their very nature and size pose greater environmental threats if implemented than others. In addition, the amount of information needed to adequately assess potential impacts can be very greatly depending on project type size and location. Lastly, the human and financial resources needed for carrying out EIA´s are often limited and assessments can only be performed for the "most important projects".

In order to come up with our list of "most important aid project types", we looked first at how OECD countries had tackled this problem regarding their domestic activities. We found that, where EIA systems exist, the types of projects covered are determined by a list, through the use of screening criteria, or a combination of the two. Some countries for example have established a positive list of specific project types which must always be submitted to environmental assessment. (This is the approach we have taken in the Netherlands). Lists, however, can also be negative, made up of project types which need never be assessed. (A number of US federal agencies have taken this approach under NEPA by creating what are called "categorical exclusions").

Other countries have established screening criteria or guidelines which are applied to projects on a case-by-case basis to determine which ones should undergo an assessment. The number and type of criteria vary from country to country but they usually attempt to determine the significance of such issues as:

- changes in the natural, physical or social environment,
- pollution,
- cumulative effects,
- endangered species,
- sensitive ecosystems,
- level of controversy, etc.

(Both Canada and Australia apply such screening criteria in deciding whether or not to carry out an EIA on a particular project).

In addition to the types of projects covered by the EIA systems of OECD Member countries, we also looked at the steps taken by various developing countries to identify project and programme types in need of EIA. Lastly we investigated types of projects which have actually been assessed on ad hoc basis by both bi-lateral and multi-lateral lending institutions.

The result of this research and the corresponding discussions within our group led to an agreement on seven types of aid projects and programmes most in need of EIA. They are:

(a) substantial changes in renewable resource use (e.g., conversion of land to agricultural production, to forestry or to pasture land, rural development, timber production);
(b) substantial changes in farming and fishing practices (e.g., introduction of new crops, large scale mechanization); use of chemicals in agriculture (e.g., pesticides, fertilizers);
(c) exploitation of hydraulic resources (e.g., dams, irrigation and drainage projects, water and basin management, water supply);
(d) infrastructure (e.g., roads, bridges, airports, harbours, transmission lines, pipelines, railways);
(e) industrial activities (e.g., metallurgical plants, wood processing plants, chemical plants, power plants, cement plants, refinery and petrochemical plants, agro-industries);
(f) extractive industries (e.g., mining, quarrying, extraction of peat, oil and gas);
(g) waste management and disposal (e.g., sewerage systems and treatment plants, waste landfills, treatment plants for household waste and for hazardous waste).

This list of project types, together with issues which should be considered when carrying out EIA´s (e.g. the effects on certain types of areas which should be assessed) make up the first recommendation of the group. The "Recommendation of the OECD Council on the Environmental Assessment of Development Assistance Projects and Programmes" was officially adopted by the Environment Committee at Ministerial level on June 18th, 1985.

In our work on the second objective - identifying the constraints faced by developing countries in carrying out environmental assessment - we were able to call upon the personal experience of many of our members in conducting assessments in the Third World. In addition we also undertook a general literature search and we contacted the governments of a number of developing countries directly. The countries for which such information was gathered included Brazil, Ecuador, Indonesia, Kenya, Malaysia, Mexico, Rwanda, Somalia, Sri Lanka, Sudan, Tanzania and Thailand. The constraints we identified were:

(a) Lack of political will and/or awareness of the need for environmental assessment;
(b) Insufficient public participation;
(c) Lack of an institutional base;
(d) Insufficient skilled manpower; and
(e) Insufficient financial resources.

The extent to which any or all of these constraints are operative in the developing world varies from region to region and country to country. For example, we found that, generally speaking, each of these constraints would apply to most African countries. In South-East Asia, on the other hand, "insufficient data" and "skilled manpower" are, for the most part, bigger constraints than legislative or institutional frameworks which are already in place.

On the whole we found that in many ways, South-East Asia and the Pacific was the most advanced region in the developing world regarding the establishment of and experience with environmental assessment. A recent study revealed, for example, that 66% of the countries in that region have passed legislation requiring EIA for certain types of projects as compared to 57% of Latin American countries and 41% of the countries in Africa and the Middle East.

Developing countries on their own and together with help from OECD Member countries are beginning to take steps to overcome these constraints. To mention a few examples:

- The Canadian government has initiated an "Environmental Manpower Development" project in Indonesia to help that country overcome the lack of trained individuals to undertake environmental assessment. Through a combination of training courses at Indonesian universities and environmental centers together with a programme to bring Indonesians to Canadian universities, a large number of government officials, academics and local consultant firms are acquiring the necessary skills for conducting EIA's.

- The U.S. Agency for International Development together with UNESCO's Man and the Biosphere Programme has completed over 40 "environmental profiles" of developibg countries. These reports are the first step in a process to develop better information for AID missions, host country officials and others on the environmental situation in specific countries and begins to identify the most critical areas of concern.

- Although not widespread, there is also growing evidence that, where aid agencies do carry oud an EIA, they are involving host country officials and others in scoping sessions, the collec-tion of base line data and the actual preparation of assessments.

One third objective was to look at the experience to date of OECD Member country aid agencies in actually carrying out environmental assessments.

Nine Delegations (Australia, Canada, Denmark, France, The Netherlands, Norway, Japan, the United Kingdom and the United States) prepared a total of 16 case studies on environmental assessments which had been carried out by their aid agencies in Asian, African and Latin American countries.

As most of these agencies do not have a legal requirement to conduct EIA's, the cases represent for the most part, ad hoc or special situations in which an assessment was carried out because of the obvious potential environmental effects or because the host country government specifically requested it.

The case studies cover six of the seven types of projects and programmes identified under objective 1 as being "most in need of environmental assessment".

Examples include highway projects in Peru, Sudan and the Phillipines, hydro-electric power projects in Sri Lanka and Indonesia, an irrigation project in Cameroon and a coal fired power station and steel mill in Singapore.

In analyzing the case studies we looked at 4 main issues:

(1) the form of the EIA;
(2) the way the EIA was prepared (i.e. who was responsible for its preparation? How did they do it?)
(3) the content of the EIA, including
  - the description of the existing environment,
  - the number and types of alternatives considered,
  - the identification and assessment of environmental impacts, and
  - mitigation measures;
(4) the effect of the EIA on project/programme outcomes.

Under the first point, form of the EIA, the case studies showed that two general approaches have been taken. Either a special environmental impact statement or report was prepared or information on the environmental aspects of the project/programme was incorporated as a part of the feasibility study or other project planning documents.

We did not find a link between the form of the assessment and its content or effect on project decisions taken. In other words, there was no apparant advantage of one form over another in terms of assessing environmental impacts or aiding the decision.

Regarding the second subject, the way in which the EIA was prepared, in approximately half the cases the documentation was prepared by outside consultants - either individuals or consultant firms - and in the other half by aid agency officials. Whether prepared "in-house" or "outside", where an inter-disciplinary team was used the resultant EIA was more comprehensive than when one individual undertook the task alone. Regardless of who actually prepared the assessment, all case studies showed evidence of coordination between those responsible for preparing the assessment and host country officials. Indeed, in some cases it was the host country government itself which requested the aid agency to undertake an assessment. This was a particularly interesting finding as it contradicts a widely held attitude in "development aid circles" that the biggest constraint to carrying out EIA in developing countries is the attitude of the countries themselves.

Under the third point, the content of the EIA, the projects described in the case studies differed so widely that it is difficult to make any generalized statements or conclusions.

For example, the extent to which the existing environment was described in the case studies varied greatly. For some projects extensive descriptions of the natural and man-made environmental setting were provided. For others, only those environmental factors which would be directly affected by the project, such as air and water quality, were described.

Regarding alternatives, most of the case studies concerned a single, site-specific development proposal without any indication that alternatives - either alternative sites or alternative means for achieving development - were considered. In such cases the EIA served the purpose of a mitigation plan for limiting negative impacts rather than a decision-making tool for comparing alternative means of achieving development goals. In some cases, however, particularly those where highway construction was involved, alternatives <u>were</u> assessed and compared on the basis of their potential environmental impacts.

The actual number and type of environmental impacts which were assessed seemed to be dependent upon the characteristics of the project itself and the way in which the EIA preparer viewed the situation. In other words, we found very little evidence that EIA guidelines or checklists related to the project type were used. In two cases a scoping process took place to identify the most significant impacts. In one of these cases - a proposal to construct electrical transmission lines on the Caribbean island of Dominica - the scoping meeting took place between USAID's regional environmental management specialist and cabinet level officials including the Prime Minister.

Almost all the case studies showed evidence of the adoption of measures to mitigate the negative environmental impacts associated with the project or programme. In some cases the mitigation was in the form of project design changes such as the inclusion of waste-water treatment facilities in a steel production facility (Singapore). In other cases it was in the form of a mitigation programme such as the creation of a wildlife preserve to compensate for the conversion of wildlife areas to agricultural production (Sri Lanka).

In no case did the environmental assessment result in a decision to halt the project or the programme. In several cases, however, the assessment brought about project or programme design changes for reducing negative environmental impact. A hydroelectric project in Indonesia, for example, included design features to mitigate the impact on the natural environment and tribal life styles, while maintaining adequate energy production.

As a general, concluding observation it might be said that, in comparison to assessments of similar projects assessed under domestic EIA requirements (such as in the USA and in Canada) the content of the aid EIA's was not as comprehensive. The factors responsible for a successful assessment, however, appear to be the same.

Based on the case studies, we have identified five "key elements" for a successful EIA and they would seem to apply wherever EIA is applied. They are timing, personnel, scoping, information and monitoring.

(1) Timing. All the case studies pointed to the need to integrate environmental assessment at an

early stage of project planning. Where environmental assessment is seen as an "extra" or "add-on" to projects which have already been determined on the basis of their engineering, technical and economic feasibility, it can perhaps suggest mitigation measures but can have no effect on project design. Where integrated early in project planning it can result in projects with built-in mitigation which are designed to minimize negative elements and maximize benefits.

(2) Personnel. The success of an environmental assessment is very much dependent on the individual (or team of individuals) responsible for preparing it. Because of the large diversity of project and program types to which assessment has been and can be applied, it is difficult to determine an ideal profile for an "EIA-preparer" which would fit every situation. As our case studies point out, some types of projects can be adequately assessed by a single person with the right qualifications and experience working together with host government officials over a short amount of time. Other projects demand interdisciplinary teams of experts to carry out extensive field investigations and data gathering. In both cases the need could conceivably be met within aid agencies themselves. A more likely situation, however, is one in which the "developer" will have to approach private consultants or consultant firms for help. In those situations, it is necessary that terms of reference be prepared in such a way as to ensure that the individual or group brings sound environmental knowledge and experience to the job.

(3) Scoping. A crucial task in carrying out environmental assessment is the determination, early in project planning, of the most significant (i.e. serious) environmental impacts associated with a project and the reasonable alternatives availa-

ble for constructing the project in the most environmentally sound manner. Scoping is a procedure for accomplishing those tasks. An early meeting of the donor agency, host government officials, environmental experts and other interested parties to "scope" the project can result in quicker, less expensive and more efficient environmental assessments.

(4) Information. The need for reliable scientific data and information is a common theme of our case studies. Where an adequate data base is missing it becomes particularly important to work closely with local universities, research institutes and the effected public to obtain an insight into existing environmental conditions. Because of the time and expense involved in "starting from scratch" it is advisable to tie data gathering to the major environmental impacts identified during scoping.

(5) Monitoring. One of the most important lessons to be learned from experience with environmental assessment is the need for monitoring. Although it is not required by any aid agency, most if not all are coming to see the need for auditing completed projects not only as a sound management measure but also in order to test the accuracy of the environmental assessments. The knowledge, for example, of which impacts proved significant and which did not can result in the improved scoping of future projects.

Having completed the work on our first three objectives, the group has now begun addressing the fourth objective on the kinds of processes, procedures, organisation and resources needed for the environmental assessment of development projects and programmes.

Although we have not yet decided on the specific recommendations we will make, the general thrust of the necessary steps we think need to be taken is clear. For example, the results of our work so far indicate that EIA needs to be seen as a comprehensive process integrated at an early stage of project and programme planning, coordinated with the host country government reflected in the implementation of the activity and followed up by monitoring and post-audit evaluation.

Though some aid agencies are more enlightened than others in the way they view environmental assessment, in none of them at present is this comprehensive approach in operation. It is of course, not necessary that all aid agencies adopt identical processes and procedures for EIA. We do feel, however, that a number of common goals and principles should be agreed upon and we will be formulating these in the coming months.

One step we will definitely be taking is to urge aid agencies to adopt a specific, clearly-stated policy related to EIA. Only through a specific policy can implementing officials be given clear directions on carrying out environmental assessment. When followed up by a commitment of skilled personnel and money we will hopefully begin to see more evidence of development which ensures economic growth while preventing environmental degradation and protecting the long-term productivity of the natural resources on which development is based.

# A NEW FORM OF INDUSTRIALIZATION: MICROCOMPUTERS IN DEVELOPING COUNTRIES

Alan L. Porter*
Frederick A Rossini*
Hengde Shi**

## ABSTRACT

The Information Revolution is unlike the Industrial Revolution. Information technologies, epitomized by the microcomputer, differ from the technologies of industrialization. These new technologies, by virtue of their accessibility, are appropriate to rapid, though uneven, assimilation by developing countries. Simple benefit/cost estimations demonstrate the economic attraction of adoption of microcomputers. However, the impacts of such adoption include expansion of home markets and increased technical employment, but also overall higher underemployment and forced social change. Impact profiles reflect a nations's current stage of development and a number of specific factors identified through a technological delivery system model. This model offers a planning tool for the Information Revolution.

## KEY WORDS

Information revolution, information technologies, microcomputer, assimilation by developing countries

---

* Georgia Institute of Technology, Atlanta, Georgia 30332

* Georgia Institute of Technology, Atlanta, Georgia 30332

** Scientech Information Center, Ministry of Machine Building Technology, Beijing, China

The industrialized nations are in the midst of the Information Revolution. We feel the impact of computers and communication technologies at work and at play. Considering that about half of the production work is really hands-on labour, information work now constitutes some 86 percent of the work force in the U.S. (Mundel, 1983). From video games to airline reservations, "compunications" affect our lives.

This Information Revolution is also reaching into the developing countries. This paper focuses on the microcomputer as a key component of this new technology and its implications for developing nations.

Some 130 developing countries contain over half the world's roughly 5 billion people (Mahadeva and Thillainayagam, 1984). They range from nations on a pre-industrial level (e.g. certain African countries concerned with agricultural production and public health) to those on a second, beginning industrialization level (e.g., certain Latin American countries), and on to those on a third relatively advanced industrialization level (e.g., certain Asian nations heavily involved with technological adoption and industrialization operations)(East, 1983). The implications of microcomputers and other information technologies vary according to economic development and also natural resources, political structures, and technological capabilities. Whatever the particulars, this vast market has drawn the interest of those who produce information technologies.

Computer capabilities continue to become ever more attractive as they build on exponential rates of improvement in component technology to:

* Double capacity every 1.5 years (e.g., memory; function/volume)
* Double speed every 1.5 years (e.g., instructions executed, sec);
* Cut costs in half every 2.5 years (e.g., of logic elements and of random access memory)(Porter, 1984).

Microcomputers, thus, provice increasing capability in a form much easier to afford and to maintain than earlier generations of mainframe computers. The next section offers a framework to address the implications of microcomputers (and, implicitly, of information technologies in general) for the developing countries.

## A technological delivery system

Figure 1 maps the institutional actors and basic interactions involved in applying microcomputers in developing countries. The horizontal axis depicts the coming together of hardware, software and networks, that constitute the technologies in question, to be applied by a technological enterprise to achieve certain payoffs, with consequent impact. Impinging on this enterprise are various societal influences, primarily acting through government entities. Let us work our way along the technological enterprise, then feed back down through the government influences.

The hardware is ubiquitous. American microcomputer sales, growing 44 percent annually, are expected to reach $25 billion by 1987 with 24 million in use (Kenealy, 1983). We project a microprocessor-based keyboard device to be in service for every white collar worker in the U.S. by 1990 (Porter et al., 1985). Microcomputers are sufficiently robust to work well in the harsher conditions of humidity and temperature typical of so many developing countries (Skorov, 1978). The developing countries also do not lack information on these mass market technologies. Computer manufacturers, especially Japanese and European, are directing efforts to sell computer to Africa, Asia, and Latin America. Worldwide desktop computer shipments are projected at 10.7 million units in 1986, a 66 percent annual growth rate in sales (Computerworld, 1983).

An alternative, of developing countries building their own microcomputers, is becoming more feasible.

Designs can be borrowed. Components are available across widely competitive markets, and themselves frequently manufactured "off shore" in developing countries to take advantage of lower material and labour costs (Rothschild, 1984). Efforts such as IBM's large scale manufacturing of personal computers in Mexico and like developments in Malaysia, promise to develop local suppliers, skilled production workers, and indigenous engineers (Wall Street Journal, 1984; Dhanan, 1984).

Software poses a similar choice: to import or to produce indigenously. Either way, the intent is to adapt operating systems and application programs to local needs. Again, outside suppliers are interested in selling their wares (e.g., NCR provides a Chinese character computer - aeu, 1983). Companies investing in such specialised development may seek exclusive licences. Developing countries may prefer to write their own software. Start-up costs appear smaller and technical skills more readily gathered than for their own hardware development. For example, India has brought out computers using Tamil and Hindi (Mahadeva and Thillainayagam, 1984).

In any event, software programs are becoming more user-friendly. That means less technical sophistication is demanded of users, lowering a traditional barrier to technology adoption by developing countries.

Networking further boosts microcomputer capabilities. Multi-user systems can employ multi-tasking, high performance microprocessors and share disk storage, printers, and software resources. They can facilitate office automation, process control (as in a chemical plant), and management control. Local area networks may be particularly important where the telephone system is not adequate for precise data transmission. International data transfer and electronic connection standards, as well as satellite networks, are also under development (East, 1983). Developing countries cannot usually afford to install a complete network at once. Therefore, it is vital that their microcomputers use compatible protocols to enable later network development (Jasany, 1984).

Figure 1 proceeds to suggest that these technologies can be used across a full range of economic sectors. Whether they are depends on availability of capital, firm size, level of competition, resistance from labour, and the degree to which computers can alter the end products provided by the sectors. Even in labour-intensive banking in a developing country, computers could improve services so much that they would be desirable. The microcomputer can find application in offices across a number of sectors, initially on a personal use basis and, eventually, through integrated systems.

Manufacturing uses microcomputers in additional ways. In particular, process industries rely on computers for control. Microcomputers make distributed control possible, and their simplicity and low cost are attractive. Individual controllers can be utilized at first, later to be linked in distributed control systems. Agriculture offers another priority sector in certain developing countries. Centrally located microcomputers-facilities can provice information to area farmers on planting, fertilizing, pest control and harvesting. Education is especially important. Teachers and students need to be trained in computer skills, and certain workers should be retrained.

A number of infrastructures are necessary to implement these technological innovations. First of all, capital is required. Government or special vendor arrangements may provide alternatives to traditional commercial sources. With declining microcomputer prices, shortage of funds is not as serious a problem as other infrastructure inadequacies (Nwachukwu, 1983). Trained manpower may be especially short in societies only beginning to industrialize. Institutionalization of technical education and training are critical to graft information technology onto such pre-industrial societies. Such training could destabilize a society if a distinct, technologically sophisticated group coexisted with a suspicious, unsophisticated one. Recall the example of

Iran under the Shah where a high tech veneer was grafted unto a fundamentalist, Moslem society with revolutionary results.

Instrumentation and equipment should not be underestimated. For instance, production automation requires not only microcomputers and programmable controllers, but also controllable machines and specialized instrumentation that may be the dominant cost factors. Development of adequate supplies, implementation of complex systems, and maintenance may dwarf problems with the microcomputers themselves. Problems may be compounded by lack of reliability of phone data transmission and electrical service. Above all, management must come to appreciate the payoffs, not feel threathened by the changes, direct the innovation process, and secure commitment of employees and cooperation of producers (Kawamura and Silva, 1984).

Should all these pieces fit into place, one can contemplate the payoffs. The "push" provided by the supply of powerful information technologies needs to be complemented by the "pull" of a demanding market. Industrial organizations, in fact, may implement the same technologies to accrue quite different payoffs:

* New products or services not previously available;
* Increased quality of present products or processes,
* Faster delivery of a desired product or service (reduced response time)
* Higher productivity by the same work force
* Operational savings in areas such as energy use, materials, paper, or floor space.

Usually, developing economies are labour-intensive so that productivity gains alone are not socially attractive.

Impact Assessment.

The last element in Figure 1 is the issue of impact. The information revolution promises significant improvements in the quality of life for developing countries. More and better services and goods will become available. In contrast to the products of industrialization, most of the information age outputs are far less resource-intensive. Televisions, telephones, and computers require far less energy or materials than, say, automobiles.

Developing nations should see an expansion in both home markets and international markets for information products. Production of such services and goods at attractive price increases their competitiveness in world markets, and that renders them less dependent on developed countries' economic and political power.

The major concern raised about the impact of information technologies, such as the microcomputer, is the displacement of labour. Both industrial and information technologies increase demand for highly skilled workers while diminishing overall labour needs. Skorov (1978, p. 90) cites evidence for ten developing countries that production is rising far faster than employment in each. He estimates semi- and unemployment between a quarter and a third of their labour force in the mid-1970's, and finds experts unanimous that this is increasing. One estimate projects an 8.5% annual increase in developing countries' semi- and unemployment to an unconscionable 70+% in 1990 (Skorov, 1978).

The potential counter to this desperate employment picture lies in market expansion at a rate greater than productivity gain. A relatively advanced industrialized society with a developing service sector should be positioned well to expand service employment markedly, but will experience rather severe dislocation in manufacturing as production is increasingly automated. In contrast, a beginning industrial society may have fewer industrial workers subject to dislocation, but it faces a greater challenge in initiating a viable service economy. A predominately agricultural society faces the most desperate situation as information technology calls for technically trained workers while the unneeded and untrained farm worker has nowhere to go.

During the mechanization of the industrialized west, an expanding manufacturing sector first accepted displaced farmers, then an expanding service sector took on displaced factory workers. Today, these developed countries face an unprecedented, simultaneous automation in two dominant sectors - production and information services. But the developing countries may face concurrent automation in all economic sectors.

To illustrate the issues raised by the introduction of microcomputers into developing countries, we offer three limited cost-benefit examples.

As a first example, automation of the postal check system in Algeria resulted in the following changes from 1974 to 1977 (Rada, 1983):

* Volume handled: + 38%
* Time to process documents: down to 15 to 2 days
* Waiting lines to pay for checks: down from 3-6 hours to 2 minutes
* System saturation: down from 95% to 50%
* Employment: -21%

In sum, the quality of service improved dramatically, leading to an increase in workload. However, the work load increase was more than compensated by productivity gains, resulting in a reduction of 176 jobs through attrition.

As a second illustration, we summarize the results of a cost-benefit calculation on the introduction of a multi-user, micro-based, small U.S. business system (Shaw, 1981).

| | | |
|---|---|---|
| Costs: | Initial | |
| | Hardware | $ 19.000 |

| | |
|---|---|
| Software | 6.000 |
| System design & data conversion | 8.000 |
| Training personnel | 10.000 |
| Conversion costs | 5.000 |

Annual

| | |
|---|---|
| Hardware maintenance | 2.000 |
| Programmer (to maintain/modify software) | 12.000 |

Benefits: Annual

| | |
|---|---|
| Reduced Xerox cost | $ 5.800 |
| Elimination of billing clerk | 9.600 |
| Reduced maintenance | 1.800 |
| Elimination of service bureau | 1.700 |
| Reduction of bad debt | 1.400 |
| Improved customer service (assuming, e.g., fill-rate improvements translate into sales increase) | 4.000 |
| Productivity gains taken as business expansion without staff increases | 15.000 |
| Improved marketing based on better data and analyses | 10.000 |

This illustration does not try to quantify a number of anticipated improvements (e.g., statements mailed sooner, fewer errors, management improvement possibilities). Nonetheless, for a 7 year life, 15% discount rate, and 50% tax rate, it pays back the investment in 2 years. Gains appear proportionately greater as transaction speed and volume increase, operations are interdependent, and control of operations becomes complex.

As the third example consider the use of a microcomputer to control a multiple-effect evaporator in a process industry (U.S.)(DOE).

Costs: First year

| | |
|---|---|
| Microcompuer (single board) | $ 500 |
| Peripherals | 6.900 |
| Instrumentation (sensors, etc.) | 24.215 |
| Software development | 4.000 |
| Training personnel | 10.000 |
| Engineering analysis & support | 20.000 |

Benefits: Annual

| | |
|---|---|
| More saleable product per unit of raw material using less steam | $ 88.600 |

This calculation does not address ongoing costs (e.g., maintenance) or other potential benefits (e.g., product quality improvement, labour savings, increased production). For a 7 year project life, 15% discount rate, and 50% tax rate, the investment pays back fully in 1.5 years. Note the relatively small cost of the microcomputer and the attractive economics.

In such a cost-benefit calculus, developing countries need to consider the costs of building infrastructure (Figure 1). However, infrastructure also carries indirect benefits, such as the use of training mechanisms for workers in other positions as well. Indeed, introduction of microcomputers will also enhance the productive base, increase computer literacy, encourage industrial competititveness, and build systems design experience useful for other automation efforts.

## Conclusions

We should not dispense with the technological delivery system (Figure 1) without emphasizing the important role of government. In many political systems, government operates the technological enterprise per se. In others, it brokers the balance between public and private operations, taking into account differences in the respective costs and benefits. For instance, Brazil has established an agency to design and regulate the development of information technology (Kawamura and Silva, 1984). In any event, "government holds the ultimate responsibility for the assessment of the appropriateness of technology reaching the economic and social goals of the nation" (Behrman, 1980, p. 188).

The microcomputer technological delivery system serves to identify a number of necessary elements for a developing country considering the adoption of such technologies. The opportunities afforded by the technologies (hardware, software, and possible networks) must match with desired payoffs of adoption. The technological enterprise must piece together the necessary infrastructure and financing to link these. Government has a responsibility to guide the process and to assure that the impacts are as socially beneficial as possible. Decisions must be made about whether to purchase the technology from foreign suppliers or to build it, possibly in cooperation with those foreign companies. Above all, the possible erosion of already weak employment opportunities must be addressed.

These concerns notwithstanding, the developing countries will adopt microcomputers and other information technologies. Increasing performance, increasing simplicity of operation, and decreasing costs, with good information across a worldwide marketing network, foreordain this result. Not to do so would undercut any cost advantages they have: automated production in a developed country tends to be cheaper than manual production in a developing country (Rada, 1983).

This is not to say that the Information Revolution will follow the same pattern across all developing countries. Developing countries typically adopt new technologies in a few production units or economic activities (Skorov, 1978). Even in a nation such as India with major pockets of industrial development, the percentage of the work force engaged in manufacturing and that in agriculture has changed less than 1% from 1921 through 1971 (Narisimhan, 1983). We believe the adoption of the microcomputer and related information technologies will be markedly different. Its wide accessibility will lead not to a uniform adoption by all enterprises in all sectors, but, we predict, a pockmarked moonscape - many individual high potential applications in many areas of production and office work, as well as services and agriculture. Some of these initial applications will later fuse together as networks are formed, and integrated data and control systems are planned. Many workers will be exposed to the computer. They, in turn, could extend computer literacy among the children with far reaching impacts. One possibility is that traditional and modern economic activities would be forced to confront each other on a wide scale, in a short time. The microcomputer could penetrate "the dense mass of archaic social structures", acting "as a catalyst of social change" (Skorov, 1978, p. 9). The changes starting to take place in the developing countries are original in character; they are not a replay of the Industrial Revolution. In the Industrial Revolution the stimulus for change was primarily internal; for the developing countries today it is external. Furthermore, and most significantly, the introduction of information technologies is concurrent with the introduction of mechanical technologies. In addition, this modernization will expand exployment far less than did the Industrial Revolution, because the higher technology of today is less labour-intensive and the populations are much more massive.

The developing nations have been slow to industrialize and have not reached the levels of the U.S. and Europe because of capital shortages, high cost of technology, shortfalls of trained personnel, and inadequate markets. The microcomputer radically changes this equation. The technology is cheap and easy to use, and it can induce indigenous and export market opportunities. Joint ventures may develop in particular information technologies between developing nation organizations and foreign corporations. Developing nations may seize opportunity in areas where developed nations rely on outmoded technology and pay higher labor rates. New professional jobs may expand dramatically in these nations.

The information age is here, not just in the developed nations, but spreading across the whole world. It permeates the entire spectrum of human activities - in offices, shops, factories, education, entertainment, agriculture, and international affairs (Mahadeva and Thillainayagam, 1984). It will affect economic development by adding new areas of economic activity for the developing nations. It will improve their productivity and quality in the primary and secondary production sectors, and instigate growth in the critical professionalized service sector (Narasimhan, 1983). The developing nations are embarking on a new venture in development. They need not, and cannot, retrace the development stages of Europe and the U.S.

## REFERENCES

aeu (Asian Electronics Union Journal), Februari 1983, 81.

Behrman, J.N.(1980). Industry ties with Science and Technology Policies in Developing Countries, Oelgeschlager, Gunn & Hain, Publishers, Cambridge, MA.

Computerworld, 28 March 1983, 55.

Dhanan, V. (1984). Malaysia Plans a Technology Leap, Cumput. Asia (Hong Kong), No. 65, March, 36-48.

DOE/cs/2123-T2, Computer Technology: Its Potential for Industrial Energy Conservation.

East, H. (1983). Information Technology and the Problems of Less Developed Countries, The Information Society Journal 2, 53-64.

Jasany, L.D. (1984). Tying the Factory Together with PCs and Networks, Production Engineering 31, 52-62

(April).

Kawamura, K. and W.D.S.E. Silva (1984), On Developing the Microelectronics Industry: A Systems Approach, Systems Research 1, 45-53.

Kenealy, P. (August 1983), Desk-Top Personal Computer, Mini-Micro Systems.

Mahadeva, K., and V.K. Thillainayagam (1984). Electronic Information Technology for the Lesser Developed Countries, The Radio and Electronic Engineer 54, 256-262.

Mundel, M.E. (1983). Improving Productivity and Effectiveness, Prentice-Hall, Englewood Cliffs, NJ.

Narasimhan, R. (1983). The Socioeconomic Significance of Information Technology to Developing Countries, The Information Society Journal, 2, 65-79.

Nwachukwu, E.O. (June 1983). Problems of Computer Management in a Developing Country, Computer (USA) 16, no. 6, 95-96.

Porter, A.L. (1984), Work Issues Raised by Office Automation Call for Expertise of Industrial Engineers, Industrial Engineering 16, 25-32.

Porter, A.L., J.D. Roessner, F.A. Rossini and K.R. Nelms (1985), Office Automation Outlook: 1985 - 2000, Report to the U.S. Congress, Office of Technology Assessment.

Rada, J.F. (1983). Information Technology and Developing Countries, "Systems Approach to Appropriate Technology Transfer", edited by P. Fleissner, Pergamon Press, 39-49.

Rothschild, K. (1984). Peripheral Mfrs. Making Move to SE Asia, Electronic News, 22 October.

Shaw, D.R. (1981). Your Small Business Computer, Van Nostrand Reinhold Company, Chapter 8, 9, 89-128.

Skorov, G.E. (ed.)(1978). Science, Technology and Economic Growth in Developing Countries, Pergamon Press, Oxford, England (translated by J. Warren).

Wall Street Journal, 29 October 1984, 37.

AUTHOR REGISTER

KEY WORD INDEX